COMMUNICATION NETWORKS AND COMPUTER SYSTEMS

A Tribute to Professor Erol Gelenbe

Communications and Signal Processing

Series Editors: A. Manikas *(Imperial Coll. London, UK)*
                A. G. Constantinides *(Imperial Coll. of Sci., Tech. & Med., UK)*

Vol. 1   Communication Networks and Computer Systems:
          A Tribute to Professor Erol Gelenbe
          ed. Javier A. Barria

COMMUNICATION NETWORKS AND COMPUTER SYSTEMS

A Tribute to Professor Erol Gelenbe

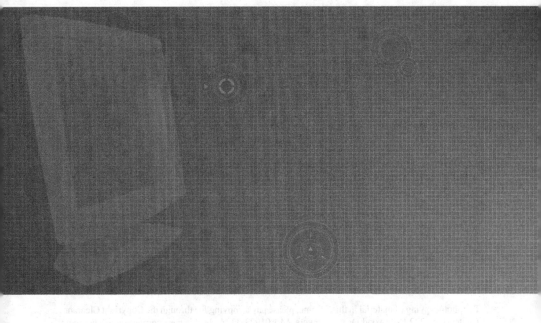

Editor

Javier A Barria
Imperial College London, UK

Imperial College Press

Published by

Imperial College Press
57 Shelton Street
Covent Garden
London WC2H 9HE

Distributed by

World Scientific Publishing Co. Pte. Ltd.
5 Toh Tuck Link, Singapore 596224
USA office: 27 Warren Street, Suite 401-402, Hackensack, NJ 07601
UK office: 57 Shelton Street, Covent Garden, London WC2H 9HE

British Library Cataloguing-in-Publication Data
A catalogue record for this book is available from the British Library.

COMMUNICATION NETWORKS AND COMPUTER SYSTEMS
A Tribute to Professor Erol Gelenbe

Copyright © 2006 by Imperial College Press

All rights reserved. This book, or parts thereof, may not be reproduced in any form or by any means, electronic or mechanical, including photocopying, recording or any information storage and retrieval system now known or to be invented, without written permission from the Publisher.

For photocopying of material in this volume, please pay a copying fee through the Copyright Clearance Center, Inc., 222 Rosewood Drive, Danvers, MA 01923, USA. In this case permission to photocopy is not required from the publisher.

ISBN-13 978-1-86094-659-2
ISBN-10 1-86094-659-3

Printed in Singapore

PREFACE

Networked systems are at the core of a wide range of human activity, including health, business, science, engineering and social interaction. It is hence not surprising that the Internet, as well as social and biological networks, have been the focus of intense research in recent years. In this context communication networks and computer systems research is entering a new phase in which many of the established models and techniques of the last twenty years are being challenged.

Industry is now introducing and proposing a wide range of new communication technologies, many of which are based on the wireless and mobile revolution, while governments and funding agencies, such as the European Union's Framework Programmes, the UK Engineering and Physical Sciences Research Council (EPSRC), the US National Science Foundation and Defence Advanced Research Projects Agency (DARPA), are supporting research on new paradigms that better exploit these opportunities to respond to the need for ubiquitous networking in both the civilian and defence sphere. At the same time, users are now becoming more aware of the usefulness of emerging services and are taking better advantage of the convergence between the Internet and the telecommunications industry. As a consequence, the research community is continuing to free itself from past intellectual constraints, so that it may fully exploit the convergence of computing and communications.

As computer and communication networks such as the Internet grow, they are naturally evolving towards decentralised and weakly cooperative environments operated by heterogeneous administrative and physical entities. In these environments, extremely large numbers of diverse mobile devices, from powerful laptops through pocket PCs to micro-devices embedded objects, will interact and connect spontaneously. Evaluating the performance of these systems constitutes a huge challenge and current network modelling research provides a set of heterogeneous tools and techniques to address such systems, rather than the systematic approach that is needed.

The discipline of network modelling and performance evaluation is thus being challenged by the increased availability of wireless and ad hoc networks with their embedded spatially dependent, time-varying, and non-stationary nature of their traffic and interconnections. In this context, novel performance evaluation and optimisation techniques must be developed which are able to take into account the uncertainties of time and space varying environments where the requests for diverse services will be made in real time, and with very different quality of service expectations.

This book is the result of seminars that touched upon this broad and challenging field. These seminars were held at a workshop on the occasion of Erol Gelenbe's inauguration as the Dennis Gabor Chair Professor at Imperial College London in April 2004. The book is divided into three parts. It begins with an introduction based upon the vote of thanks given which followed Prof. Erol Gelenbe's Inaugural Lecture, summarising his contributions to the field of Computer and Network Performance, his work as a mentor and advisor to numerous young researchers, and his other academic contributions.

In Part I of the book we address aspects of resource management in current computer and communication networks. First, we consider the impact of incentive systems in an ad hoc network as an enabler of relaying functions. This is followed by a study of resource allocation using social choice theory and welfare economics as a mechanism to mediate the allocation of resources in networks shared by multiple types of applications. Next, a discussion of the historical perspective of the Locality Principle is presented and how our understanding of locality has evolved beyond the original idea of clustering of reference. This part ends with a study of the efficient scheduling of parallel jobs on distributed processors.

In Part II new modelling paradigms and challenges are presented. This part starts with a note of caution about attempting to draw certain inferences about the nature of a stochastic process by analysing the properties of its associated steady state distributions. Next, a review of several approximation methods to solve the non-stationary loss queue problem is presented. This is followed by a review of the numerical instabilities of computational algorithms for product-form queuing systems. Stabilisation techniques for load-dependent models are discussed. The importance of understanding the behaviour of the increasingly complex network of interdependent critical infrastructure is discussed and an overview of emerging methods and tools is presented next. This is followed by a discussion of the numerical difficulties faced when solving large-scaled Markov chain models. Bounding procedures are presented as a means of circumventing these difficulties. Finally, aggregation techniques for solving large-scaled complex systems are reviewed and their suitability for a cross-layer simulation environment is discussed.

In Part III, the network modelling of emergent computer and communication networks is considered. First, the space capacity paradox and the time capacity paradox in mobile ad hoc networks are discussed and a precise quantification of these properties is provided. This is followed by a study of the benefits and possible vulnerabilities of a peer-to-peer enabled control and command structure for a time critical mission. Next, a real time mechanism that quantifies the users' subjective perception of quality of audio and video is presented. This part concludes with a study of the dynamic behaviour of the content of a popular news web site.

All the above studies highlight the scientific and technical challenges faced by current and future networks, and the need to support them through research on novel modelling and performance evaluation techniques. As we conclude this preface, let us note that enhanced network modelling techniques can lead to fast and economic service deployment and effective dynamic resource management, and that they can also enable new business strategies and infrastructures that will facilitate the emergence of future services and applications.

J.A. Barria
Imperial College London
London, U.K.

CONTENTS

Preface   v

1. Erol Gelenbe's Contributions to Computer and Networks Performance   1
Alain Bensoussan

    1.1 Introduction and Background 1
    1.2 Technical Contributions 2
    1.3 Contributions as a Research Leader and Mentor 4
    1.4 Service to the Profession 5
    References 5

Resource Management   9

2. Rethinking Incentives for Mobile Ad Hoc Networks   11
Elgan Huang, Jon Crowcroft and Ian Wassell

    2.1 Introduction 11
    2.2 Token Based Incentive Systems 12
        2.2.1 Quality of Service Problems 12
        2.2.2 Technical Conundrums 13
    2.3 Trust Management Systems 15
    2.4 Transparency vs Choice 16
    2.5 Proposed Solution 17
        2.5.1 Adoption Cycle For Mobile Ad Hoc Networks . . . 18
        2.5.2 Do We Really Need Incentive Systems? 20
    2.6 Conclusions 22
    References 23

3. Fair and Efficient Allocation of Resources in the Internet   25
Ronaldo M. Salles and Javier A. Barria

- 3.1 Introduction 25
- 3.2 Fairness, Efficiency and Utility Functions 26
- 3.3 Utility-Based Bandwidth Allocation 29
  - 3.3.1 Utility of the Aggregate 30
  - 3.3.2 Limiting Regime Approximation 32
  - 3.3.3 Offered Load Estimation 33
- 3.4 Utility-Based Admission Control 35
- 3.5 Utility-Based Scheduling 37
  - 3.5.1 Measuring Class Delays 38
- 3.6 Conclusion 40
- Acknowledgements 40
- References 41

4. The Locality Principle   43
Peter J. Denning

- 4.1 Introduction 43
- 4.2 Manifestation of a Need (1949-1965) 44
- 4.3 Discovery and Propagation of Locality Idea (1966-1980) .. 47
- 4.4 Adoption of Locality Principle (1967-present) 56
- 4.5 Modern Model of Locality: Context Awareness 57
- 4.6 Future Uses of Locality Principle 60
- References 62

5. A Simulation-Based Performance Analysis of Epoch Task Scheduling in Distributed Processors   69
Helen Karatza

- 5.1 Introduction 69
- 5.2 Model and Methodology 71
  - 5.2.1 System and Workload Models 71
  - 5.2.2 Task Routing Methods 72
  - 5.2.3 Scheduling Strategies 73
  - 5.2.4 Performance Metrics 73
  - 5.2.5 Model Implementation and Input Parameters 73
- 5.3 Simulation Results and Performance Analysis 74
  - 5.3.1 Probabilistic Routing 75
  - 5.3.2 Shortest Queue Routing 80

	5.4	Conclusions	84
	References		85

New Challenges on Modelling and Simulation   87

6. Counter Intuitive Aspects of Statistical Independence in Steady State Distributions   89
Jeffrey P. Buzen

    6.1   Introduction 89
    6.2   A System of Two Independent M/M/1 Queues 90
    6.3   A System of Two Queues in Tandem 93
    6.4   Statistical and Dynamic Independence 96
    6.5   Beyond Stochastic Modelling 98
           6.5.1   Central Role of Steady State Distributions 98
           6.5.2   Generality, Robustness and Level of Detail 100
           6.5.3   Operational Analysis 101
    6.6   Conclusions 103
    References 103

7. The Non-Stationary Loss Queue: A Survey   105
Khalid A. Alnowibet and Harry Perros

    7.1   Introduction 105
    7.2   The Simple Stationary Approximation (SSA) Method . . . 108
    7.3   The Stationary Peakedness Approximation (PK) Method . 109
    7.4   The Average Stationary Approximation (ASA) Method . . 111
    7.5   The Closure Approximation for Non-Stationary Queues . . 112
    7.6   The Pointwise Stationary Approximation (PSA) Method . 114
    7.7   The Modified Offered Load Approximation (MOL) Method 118
    7.8   The Fixed Point Approximation (FPA) Method 121
    7.9   Conclusions 123
    References 124

8. Stabilization Techniques for Load-Dependent Queuing Networks Algorithms   127
Giuliano Casale and Giuseppe Serazzi

    8.1   Introduction 127
    8.2   Preliminaries 128
           8.2.1   Numerical Exceptions 128

	8.2.2 Closed Product-Form Queuing Networks	129
8.3	Numerical Instabilities in PFQN Algorithms	130
	8.3.1 Convolution Algorithm	130
	8.3.1.1 Static and Dynamic Scaling Techniques	131
	8.3.2 Load Dependent Mean Value Analysis (MVA-LD)	133
8.4	Improved Stabilisation Techniques	134
	8.4.1 Software Stabilisation	134
	8.4.2 Stabilisation of MVA-LD with Two Customer Classes	136
	8.4.2.1 Numerical Example	137
8.5	Conclusions	140
	References	140

9. Modelling and Simulation of Interdependent Critical Infrastructure: The Road Ahead   143
Emiliano Casalicchio, Paolo Donzelli, Roberto Setola and Salvatore Tucci

9.1	Introduction	144
9.2	Modelling and Simulation of Interdependent Infrastructures	146
	9.2.1 Interdependency Analysis	146
	9.2.2 System Analysis	150
9.3	Conclusions	154
	References	157

10. Stochastic Automata Networks and Lumpable Stochastic Bounds: Bounding Availability   161
J.M. Fourneau, B. Plateau, I. Sbeity and W.J. Stewart

10.1	Introduction	162
10.2	Stochastic Automata Networks and Stochastic Bounds	163
	10.2.1 SANs and their Tensor Representations	164
	10.2.2 Stochastic Bounds: An Algorithmic Presentation	165
10.3	LIMSUB and Analysis of Transient Availability	169
10.4	Algorithm GetCol and its Complexity	172
	10.4.1 Computation of a Column	174
	10.4.1.1 Computation of the Synchronised Part C_s	175
	10.4.1.2 Computation of the Local Part C_l	176
	10.4.2 Sorting and Uniformisation	177
10.5	Example: A Resource Sharing Model Subject to Failure	177
	Acknowledgements	179
	References	179

11. Aggregation Methods for Cross-Layer Simulations 183
Monique Becker, Vincent Gauthier, André-Luc Beylot and Riadh Dhaou

  11.1 Introduction 183
  11.2 Aggregation Methods 184
  11.3 Aggregation of Markov Chains 185
    11.3.1 Introduction 185
    11.3.2 Decomposability Method of Kemeny and Snell: Theory 185
    11.3.3 Example of the Method of Kemeny and Snell 186
    11.3.4 Decomposition Method of Courtois: Theory 188
    11.3.5 Example of the Decomposition Method of Courtois 190
  11.4 Aggregation of Physical Sub-System 191
  11.5 Time-Space Aggregation 195
  11.6 Layer Aggregation 196
    11.6.1 Dynamic Simulations 197
    11.6.2 Example of Inter-Layer Design 197
    11.6.3 Network Protocols which use Inter-Layer Interactions 198
  11.7 Conclusion 202
  References 202

Modelling of Emerging Networks 205

12. Space and Time Capacity in Dense Mobile Ad Hoc Networks 207
Philippe Jacquet

  12.1 Introduction 207
  12.2 Gupta and Kumar Scaling Property 209
  12.3 Massively Dense Networks 211
    12.3.1 Tractable Case with Curved Propagation Lines . 213
    12.3.2 Practical Implementation of Shortest Path Protocol 214
  12.4 Introduction of Time Component 215
  12.5 Information Flow Tensor and Perspectives 220
  References 220

13. Stochastic Properties of Peer-to-Peer Communication Architecture in a Military Setting 223
Donald P. Gaver and Patricia A. Jacobs

  13.1 Problem Formulation 223

13.2 A Renewal Model for Blue vs Red in a Subregion 225
    13.2.1 The Visibility Detection Process 225
    13.2.2 The Model 226
    13.2.3 The Probability an EP of Size $\tilde{b}$ Attaches to the Detected RA Before it Hides 227
    13.2.4 The Probability the Detected RA is Killed Before it Hides 228
    13.2.5 Model with Additional C2 Time 228
    13.2.6 Numerical Illustration 229
13.3 Conclusions 231
References 232

14. Quantifying the Quality of Audio and Video Transmissions over the Internet: The PSQA Approach   235
Gerardo Rubino

14.1 Introduction 235
14.2 The PSQA Technology 236
14.3 The Random Neural Networks Tool 239
    14.3.1 G-networks 240
    14.3.2 Feedforward 3-layer G-networks 242
    14.3.3 Learning 243
    14.3.4 Sensitivity Analysis 245
14.4 Applications 246
14.5 Conclusions 249
References 249

15. A Study of the Dynamic Behaviour of a Web Site   251
Maria Carla Calzarossa and Daniele Tessera

15.1 Introduction 251
15.2 Data Collection 253
15.3 Results 254
15.4 Conclusions 259
Acknowledgements 260
References 260

CHAPTER 1

Erol Gelenbe's Contributions to Computer and Networks Performance

Alain Bensoussan

Former President of INRIA and of the European Space Agency
Member of the Academia Europaea and of the Académie des Sciences de Paris

This note, which is based on a speech given by the author on the occasion of Erol Gelenbe's inauguration as the Dennis Gabor Chair Professor at Imperial College London in April 2004, describes his manifold contributions to the performance evaluation field and the scientific community.

1.1. Introduction and Background

Born in Istanbul where a Lycée and a neighbourhood carry the family name, and a direct descendant of the 18th century Ottoman cleric and mathematician **Ismail Gelenbevî Efendi (1730-1790)** who taught at the Imperial War College in Istanbul, Erol entered the Middle East Technical University in Ankara as 4th ranked, and graduated as 2nd ranked in 1966. In his first year he hesitated to switch from engineering to pure mathematics, but he ended up staying in Engineering, and received a BS with High Honours in Electrical Engineering and the Clarke Award for Research Excellence for his Bachelor's thesis on Partial Flux Switching Memory. After receiving his MS and PhD in three years as a Fulbright and NATO Science Fellow at Brooklyn Poly in New York, and publishing several papers on automata and language theory in journals such as *Information and Control* and the *IEEE Trans. on Computers*, he briefly joined the University of Michigan in Ann Arbor and then left the US because of the foreign residency requirements of the fellowships he had held. He was invited by IRIA to spend six months in 1972, before he was to join the University of Liège in Belgium as a very young chaired professor in 1974. The attraction of Paris proved too strong, and under the influence of my late teacher Jacques Louis Lions, the next twenty years were spent in Paris, at INRIA as well as teaching at the University and at Ecole Polytechnique.

From the early 70s to today, working first at INRIA and in France, and then later in the US and now the UK, Erol has made seminal contributions to computer systems and networks, and in stochastic modelling. He has published four books which have appeared in English, French, Japanese and Korean. His numerous students are now senior researchers in industry and academia, in leading institutions in France, Italy, Belgium, Greece, Turkey, Canada and the US.

1.2. Technical Contributions

Internationally recognised as a founder of his field, Erol's work contains many "firsts" in tackling computer and network performance problems and developing new stochastic models and their solutions.

In his initial work on performance modelling in the early 70s, he developed, concurrently with Frank King of IBM Research, a finite state model of paged memory management[1]. Erol proved the surprising result that the FIFO paging algorithm is strictly equivalent to a random page replacement choice for the independent reference model of program behaviour, for all possible page reference frequencies[1]. He also established a hierarchy of memory management policies, going from random choice to optimal, and derived the page fault ratio in explicit form. This approach was later taken up by leading figures in the analysis of algorithms such as Flajolet, Hofri and Szpankowski, and extended to many other algorithms.

In the mid-70s he made two seminal contributions. Concurrently with H. Kobayashi, he derived the first diffusion approximations for queuing systems but suggested a model with holding times at the boundaries providing better accuracy at light traffic, and good accuracy at heavy traffic[2]. This work was used later[3,4] and in a fairly recent IBM Patent[5] on call admission control for ATM networks. Secondly, together with G. Fayolle and J. Labetoulle, he showed that the random access ALOHA channel can only be stabilised by a control policy in which the retransmission time is proportional to the number of blocked transmissions[6–8]. This work was later extended by researchers, both the in analysis of algorithms and communications theory (Hajek, Gerla, Gurcan, Szpankowski, Hofri). In the late 70s Erol and some students built the "first ever" random access network on a fibre optic channel (the Xantos system) using these results.

In the late 70s Erol proved[9] a fundamental result on reliable database operation[10]: in a database subject to *random* failures, the optimum checkpoint interval is a *deterministic* quantity expressed in terms of the amount of work that the database has accomplished since the most recent checkpoint. He also derived an analytical expression for the value of the optimum, in terms of the transaction processing rate, the failure rate, and the cost of carrying out a checkpoint. This work later gave rise to applications in control theory (for instance, in Maurice Robin's thesis on impulse control), and I followed this with interest, and further developed in several theses by Erol's students (Francois Baccelli, Marisela Hernandez)[11,13,14] and in the US[12] (David Finkel, Satish Tripathi). Also during that time, together with his students J. Labetoulle and G. Pujolle he published some of the very first performance analyses of communication protocols.

The Gelenbe-Mitrani[15] graduate level book that appeared in 1980, and which was translated into Japanese in 1988, is a classic text in computer performance analysis that has been used at many universities, including MIT. The early and mid-80s saw him make other pioneering contributions: (a) the classical Gelenbe-Iasnogorodski[16] result characterising, under general arrival and service conditions, the long run waiting time at a paging disk server, (b) the first ever work on the analytical estimation of the space and time cost of relational database operations (with his student D. Gardy)[17] which was also presented at VLDB'82 (and which gave rise to subsequent work in the community on analysis of algorithms), and the paper (with his students F. Baccelli and B. Plateau) on the end-to-end performance of the packet re-sequencing protocol[18]. A Chesnais-Gelenbe-Mitrani paper[19] was the first model and analysis of the performance of interfering reads and writes during parallel access to shared data, and the first instance of the use of "mean field techniques" in performance modelling. In the late 80s the Gelenbe-Pujolle book[20,21] appeared in French and was soon translated into English for publication by Wiley. This book has gone through several editions (1998, 1999) and was published in Korean in the year 2000.

Over the last decade, Erol's invention of work removing "negative customers"[22] has produced a resurgence of research on "product form networks", a subject that had remained relatively inactive since the major results of the 70s and early 80s. His papers on "G-networks", which have appeared in applied probability journals and in performance evaluation conferences[18-26] have opened a new chapter in queuing network theory. This highly innovative work which is motivated by the need to introduce control functions in queuing systems, introduces several new customer types: "negative" customers represent decisions to destroy traffic, "triggers" represent decisions to move customers in the network, and "resets" represent customers that can replenish the contents of an idle queue. For each of these models, he showed product form. An interesting mathematical twist is that G-networks have non-linear traffic equations, while the traditional Jackson networks which are a special case, and BCMP networks, have linear traffic equations. This non-linearity raises questions of existence and uniqueness of solutions to G-networks. Erol investigated the use of homotopy theory in this context, and then developed the framework to prove the existence and uniqueness of the solutions to G-networks whenever the traffic intensity of each queue is less than one.

His technical contributions are not limited to these salient points. His work on G-networks has had a parallel development in the field of neural computation, where "negative customers" are models of the inhibitory spikes in biophysical neurons[27-29,31-45]. Erol has developed related learning algorithms and his group has come up with many novel applications. In the last ten years he has established an enviable reputation in the area of neural computation, and together with T. Kohonen and S.I. Amari, he was a keynote speaker at the ICANN'03/ICONIP (joint Asian/European) conference on neural networks. Erol's neural networks research is also being brought back to its source, impacting performance analysis: Gerardo Rubino's group at IRISA (France) applies Erol's neural models to evaluate the user's perceived quality of audio and video traffic.

Erol is currently conducting innovative experimental work on computer network design[46,47,49], using performance and quality of service as the driving element. His group has designed and built an experimental "self-routing packet" test-bed based on his Cognitive Packet Network patent that has won him invitations to deliver keynote lectures in 2002 and 2003 at six US and European conferences. Related to this, Erol is also developing a theory of "sensible"[48] network routing. This gives a flavour of the creativity of Erol, of his talent and of his capacity to solve hard problems. But Erol is also a Manager and an Entrepreneur, the author of several US and European patents, and a capable fund raiser.

1.3. Contributions as a Research Leader and Mentor

In the early 70s Erol founded a research laboratory at INRIA (France), and then also at the University of Paris in the 80s. These were undeniably the best such groups in Europe at the time. Many international researchers were in fact introduced to the field of performance evaluation through their contacts or visits with Erol's activities. Erol developed the first graduate program in computer and network performance in France and revitalised performance evaluation research within France-Telecom. Since then, these efforts have given rise to many groups in the Paris area (INRIA, Paris VI, ENS, Versailles, INT, France Telecom) as well as in Sophia Antipolis and Grenoble, and some of them are still among the best in the world. Current research at four of INRIA's centres, Rocquencourt, Sophia Antipolis, Rennes and Grenoble bears the mark of Erol's influence.

Erol has a remarkable track record at attracting and mentoring PhD students and younger researchers, many of whom have become technical and organisational leaders. He deploys considerable effort to secure the intellectual and material resources to support graduate students, and has graduated over 50 PhDs. He may also be commended for having graduated numerous women PhDs, many of whom are well known academics.

1.4. Service to the Profession

Finally, Erol has made seminal contributions to the organization of his field by being a founding member of ACM SIGMETRICS, and the founder (together with Paul Green of IBM) of IFIP WG7.3. A founder of the journal *Performance Evaluation*, he has been active in editorial boards such as *Acta Informatica*, *IEEE Trans. on Software Engineering.*, *Telecommunication Systems*, the *Computer Journal* of the British Computer Society, *Recherche Opérationnelle* and *Annales des Télécommunications* (France), and the *Proceedings of the IEEE*. He has served on the programme committee as a conference or programme committee chair of numerous IEEE, ACM, and IFIP conferences in Europe and the US. He has also founded and helped organise for twenty years the International Symposia on Computer and Information Sciences series in Turkey.

References

1. Gelenbe E., "A unified approach to the evaluation of a class of replacement algorithms", *IEEE Trans. Computers*, **22**(6), (1973), 611-618.
2. Gelenbe E., "On approximate computer system models", *Journal ACM*, **22**(2), (1975), 261-269.
3. Gelenbe E., Mang X., Onvural R., "Diffusion based call admission control in ATM", *Performance Evaluation*, **27&28**, (1996), 411-436.
4. Gelenbe E., Mang X., Onvural R., "Bandwidth allocation and call admission control in high speed networks", *IEEE Communications Magazine*, **35**(5), (1997), 122-129.

5. Marin G.A., Mang X., Gelenbe E., Onvural R., U.S. Patent No. 6,222,824., 2001.
6. Fayolle G., Gelenbe E., Labetoulle J., "Stability and optimal control of the packet switching broadcast channel", *Journal ACM*, **24**(3), (1977), 375-386.
7. Banh-Tri-An, Gelenbe E., "Amélioration d'un canal de diffusion sous contrôle", *Revue d'Informatique (R.A.I.R.O)*, **11**(3),(1977), 301-321.
8. Gelenbe E., "On the optimum checkpoint interval", *Journal ACM*, **26**(2), (1979), 259-270.
9. Gelenbe E., Tripathi S., Finkel D., "Performance and reliability of a large distributed system", *Acta Informatica*, **23**, (1986), 643-655.
10. Gelenbe E., Hernandez M., "Optimum checkpoints with time dependent failures", *Acta Informatica*, **27**, (1990), 519-531.
11. Gelenbe E., Hernandez M., "Virus tests to maximize availability of software systems", *Theoretical Computer Science*, **125**, (1994), 131-147.
12. Gelenbe E., Mitrani I., *Analysis and Sythnesis of Computer Systems*, (Academic Press, New York & London), 1980.
13. Gelenbe E., Iasnogorodski R., "A queue with server of walking type", *Annales de l'Institut Henri Poincaré, Série B*, **16**(1), (1980), 63-73.
14. Gelenbe E., Gardy D., "On the size of projections I", *Information Processing Letters*, **14**(1), (1982), 18-21.
15. Baccelli F., Gelenbe E., Plateau B., "An end-to-end approach to the resequencing problem", *Journal ACM*, **31**(3), (1984), 474-485.
16. Chesnais A., Gelenbe E., Mitrani I., "On the modelling of parallel access to shared data", *Comm. ACM*, **26**(3), (1983), 196-202.
17. Gelenbe E., Pujolle G., *Introduction to Networks of Queues*, 2nd Edition, (John Wiley Ltd, New York & Chichester), 1998.
18. Gelenbe E., "Queueing networks with negative and positive customers", *Journal of Applied Probability*, **28**, (1991), 656-663.
19. Gelenbe E., Glynn P., Sigman K., "Queues with negative arrivals", *Journal of Applied Probability*, **28**, (1991), 245-250.
20. Gelenbe E., Schassberger M., "Stability of product form G-Networks", *Probability in the Engineering and Informational Sciences*, **6**, (1992), 271-276.
21. Gelenbe E., "G-networks with instantaneous customer movement", *Journal of Applied Probability*, **30**(3), (1993), 742-748.
22. Gelenbe E., "G-Networks with signals and batch removal", *Probability in the Engineering and Informational Sciences*, **7**, (1993), 335-342.
23. Gelenbe E., "G-networks: An unifying model for queueing networks and neural networks," *Annals of Operations Research*, **48**(1-4), (1994), 433-461.
24. Fourneau J.M., Gelenbe E., Suros R., "G-networks with multiple classes of positive and negative customers," *Theoretical Computer Science*, **155**, (1996), 141-156.
25. Gelenbe E., Labed A., "G-networks with multiple classes of signals and positive customers", *European Journal of Operations Research*, **108**(2), (1998), 293-305.

26. Gelenbe E., Shachnai H., "On G-networks and resource allocation in multimedia systems", *European Journal of Operational Research*, **126**(2), (2000), 308-318.
27. Gelenbe E., Fourneau J.M., "G-Networks with resets", *Performance Evaluation*, **49**, (2002), 179-1922.
28. Gelenbe E., "Reseaux stochastiques ouverts avec clients negatifs et positifs, et reseaux neuronaux", *Comptes-Rendus Acad. Sci. Paris, t. 309, Serie II*, (1989), 979-982.
29. Gelenbe E., "Random neural networks with positive and negative signals and product form solution", *Neural Computation*, **1**(4), (1989), 502-510.
30. Gelenbe E., "Stable random neural networks", *Neural Computation*, **2**(2), (1990), 239-247.
31. Gelenbe E., "Distributed associative memory and the computation of membership functions", *Information Sciences*, **57-58**, (1981), 171-180.
32. Gelenbe E., "Une generalisation probabiliste du probleme SAT'", *Comptes-Rendus Acad. Sci., t 313, Serie II*, (1992), 339-342.
33. Atalay V., Gelenbe E., "Parallel algorithm for colour texture generation using the random neural network model", *International Journal of Pattern Recognition and Artificial Intelligence*, **6**(2&3), (1992), 437-446.
34. Gelenbe E., "Learning in the recurrent random network", *Neural Computation*, **5**, (1993), 154-164.
35. Gelenbe E., Koubi V., Pekergin F., "Dynamical random neural approach to the traveling salesman problem," *ELEKTRIK*, **2**(2), (1994), 1-10.
36. Gelenbe E., Cramer C., Sungur M., Gelenbe P. "Traffic and video quality in adaptive neural compression", *Multimedia Systems*, **4**, (1996), 357-369.
37. Gelenbe E., Feng T., Krishnan K.R.R., "Neural network methods for volumetric magnetic resonance imaging of the human brain," *Proceedings of the IEEE*, **84**(10), (1996), 1488-1496.
38. Gelenbe E., Ghanwani A., Srinivasan V., "Improved neural heuristics for multicast routing", *IEEE Journal of Selected Areas of Communications*, **15**(2), (1997), 147-155.
39. Gelenbe E., Harmanci K., Krolik J., "Learning neural networks for detection and classification of synchronous recurrent transient signals", *Signal Processing*, **64**(3), (1998), 233-247.
40. Bakircioglu H., Gelenbe E., Kocak T., "Image processing with the Random Neural Network model", *ELEKTRIK*, **5**(1), (1998), 65-77.
41. Gelenbe E., Cramer C., "Oscillatory corticothalamic response to somatosensory input", *Biosystems*, **48**(1-3), (1998), 67-75.
42. Gelenbe E., Mao Z.-H., Li Y.-D., "Function approximation with spiked random networks", *IEEE Trans. on Neural Networks*, **10**(1), (1999), 3-9.
43. Gelenbe E., Fourneau J.M., "Random neural networks with multiple classes of signals", *Neural Computation*, **11**(4), (1999), 953-963.
44. Cramer C., Gelenbe E., "Video quality and traffic QoS in learning-based sub-sampled and receiver-interpolated video sequences", *IEEE Journal on Selected Areas in Communications*, **18**(2), (2000), 150-167.

45. Gelenbe E., Hussain K., "Learning in the multiple class random neural network", *IEEE Trans. on Neural Networks*, **13**(6), (2002), 1257-1267.
46. Gelenbe E., Lent R., Xu Z., "Measurement and performance of a cognitive packet network", *Computer Networks*, **37**, (2001), 691-791.
47. Gelenbe E., Lent R., Xu Z., "Design and performance of cognitive packet networks", *Performance Evaluation*, **46**, (2001), 155-176.
48. Gelenbe E., "Sensible decisions based on QoS", *Computational Management Science*, **1**(1), (2004), 1-14.
49. Gelenbe E., Lent R., "Ad hoc power aware Cognitive Packet Networks", *Ad Hoc Networks Journal*, **2**(3), (2004), 205-216.

Resource Management

CHAPTER 2

Rethinking Incentives for Mobile Ad Hoc Networks

Elgan Huang, Jon Crowcroft and Ian Wassell

*University of Cambridge, Computer Laboratory, Cambridge, U.K.
E-mails: Elgan.Huang@cl.cam.ac.uk, Jon.Crowcroft@cl.cam.ac.uk,
ijw24@cam.ac.uk*

Without sufficient nodes cooperating to provide relaying functions, a mobile ad hoc network cannot function properly. Consequently various proposals have been made which provide incentives for individual users of an ad hoc mobile network to cooperate with each other. In this chapter we examine this problem and analyse the drawbacks of currently proposed incentive systems. We then argue that there may not be a need for incentive systems at all, especially in the early stages of adoption, where excessive complexity can only hurt the deployment of ad hoc networks. We look at the needs of different customer segments at each stage of the technological adoption cycle and propose that incentive systems should not be used until ad hoc networks enter mainstream markets. Even then, incentive systems should be tailored to the needs of each individual application rather than adopting a generalised approach that may be flawed or too technically demanding to be implemented in reality.

2.1. Introduction

Mobile ad hoc networks are fundamentally different from conventional infrastructure based networks in that they are self-organising and formed directly by a set of mobile nodes without relying on any established infrastructure. The network thus relies on the cooperation of individual users whose devices perform the forwarding that is necessary to achieve network capability. Without sufficient nodes providing relaying functions, the network cannot function properly.

When all the nodes of an ad hoc network belong to a single authority, e.g. a military unit or a rescue team, they have a common goal and are thus naturally motivated to cooperate. However, for general applications with large numbers of unrelated users, if battery power, bandwidth, processor clock cycles and other resources are scarce, selfish users might not wish to forward packets for other users as it would impact their own ability to transmit traffic.

These concerns have resulted in a number of efforts to design incentive systems for mobile ad hoc networks that encourage users to cooperate, as well as trust management systems that identify non-cooperating nodes and punish them. However these incentive systems have a number of inherent flaws that make them difficult and undesirable to implement in practice. Ironically, if badly implemented, some of them even have the potential to backfire by offering an incentive to cheat the incentives system in order to gain further benefits.

2.2. Token Based Incentive Systems

2.2.1. *Quality of Service Problems*

With token-based incentive systems[6,8–11,14,15], the basic idea is to use notional credit, monetary or otherwise to pay off users for the congestion costs (transmission and battery costs) they incur from forwarding packets from other users. These credits can then be used to forward their own packets through other users, resulting in an incentive to act as relay points, especially where there is the greatest excess demand for traffic since this is when they earn the most. Users who do not cooperate will not be able to use the network themselves, having not earned any credits.

This idea makes a lot of sense in theory, but when practically implemented is likely to run into a number of problems.

Under the general token mechanism, a user's token count is increased when it forwards, and decreased proportionally to the number of hops it needs when it sends. This inevitably means that a user needs to forward more than he sends and also limits the amount of information that any user can send at any given time, dependent on their store of tokens. In principle the node may be able to buffer packets until it earns enough to send, but this works only as long as the buffer is large enough and there are no delay constraints on the packets, which rules out many real time applications. Therefore, practically speaking, packets could often be dropped at the source, rendering it somewhat ineffective and inefficient for many types of communications.

The system also puts users on the outskirts of a network at a disadvantage unrelated to their willingness to participate. Those users will not have as much traffic routed through them due to their location and furthermore will have lower congestion prices because of that. They will thus earn significantly less than a node at the centre of the network. The system might indeed stabilise overall, but not at a point that is beneficial to everyone.

To pay out credit to forwarding nodes, the transmitting node must estimate the number of hops required so that it can load sufficient credit onto its packet to pay each of the nodes. This calculation not only takes up resources but if done incorrectly will result in packets that have insufficient credit being dropped, as well as wasted credit, decreasing QoS for all concerned.

Another concern is that a significant amount of energy is thus wasted in the system transmitting dropped packets that would not have been dropped had the incentives scheme not been in place. Because of the wasted energy, a user might find that his battery drained faster than if he were to cooperate with no incentives system in place, as in both cases he would be forwarding packets for others but with the incentives system he suffers additional energy loss from dropped packets.

From a general consumer's point of view, these problems collectively result in dropped packets, excessive consumption of resources and generally poor quality of service for no apparent reason, representing a rather significant drawback to the use of ad hoc devices. Users with poor quality of service are unlikely to be sympathetic (or even aware) to arguments that the system works in such a way for the greater good. This would cause problems not only for individual users, but also for the overall network as unsatisfied users leave the system completely and bad word of mouth discourages new users to join. Ad hoc networks need a critical mass of users to function well, with the utility of the network increasing proportionally to the square of the number of nodes, as stated by Metcalfe's Law[5].

2.2.2. Technical Conundrums

When using tokens, there is also the question of how the balance of tokens can be maintained for users. The average token level within the system needs to be kept at a reasonable level in order for incentives to work properly. If it grows too high, everyone will be rich in tokens and no longer have an incentive to cooperate, and conversely, if there is not enough credit within the system then hardly anyone will be able to transmit. However, if an individual's token level is regularly reset (as proposed in current systems) in order to maintain a certain token level, then there is no incentive to cooperate in the long term. Nodes are free to stop cooperating once enough credit is earned to complete their transmission, since excess credit will be lost anyway.

Some systems propose using real money as credit, either directly or indirectly[6] (to buy virtual credit). In an incentives system this could prove very dangerous, because it would in itself be a strong incentive for users to game the system in order to derive monetary gains. Unless a perfect cheat proof system can be designed, which is rather unlikely, such an incentives system would ironically make it more worthwhile for users to attempt to cheat. The need to pay to communicate would also negate one of the key advantages of ad hoc networks and make it less appealing with respect to competing technologies. Also, any system that involves real money and does not incorporate tamper proof hardware requires a centralised authority. This would undermine the self-organising, decentralised nature of ad hoc networks, as well as requiring suitable infrastructure to be built, making the networks less easily deployable and less scalable. It would also be difficult in an ad hoc network to ensure that centralised authorities would always be within coverage.

Tamper proof hardware in turn is very difficult to achieve as suggested in[3]; virtually any system can be modified. A determined hacker would be able to compromise a system regardless of whether there was a 'tamper proof' module in place (even if the module was truly tamper proof the hacker might simply replace it with one of his own design). In the end this might only discourage less technically capable users who would not have tampered with the devices in the first place.

Another problem with such systems is that it is very difficult to charge users fairly, without introducing additional complexity. In most systems presented to date it is the sender that always pays, although it is technically possible to also charge either just the destination or both. This is mainly to prevent the sender from sending useless messages and flooding the network. However, in many cases it is the destination that stands to benefit from a transmission and charging only the sender may thus lead to inconvenience to the user and thereby discourage use of the system. In the same vein, charging just the destination or even both parties would not be perfect solutions either, as the beneficiary changes with each application. (An alternative method of preventing useless messages from being sent might simply be a hardwired mechanism that throttles communications exceeding a certain rate/amount). It is also unclear how this payment issue scales to two-way communications, especially when one side has enough credit and the other does not.

Complexity of solutions is another issue. The mechanisms used to enforce these incentives systems take up resources themselves. If the proportion of freeloaders is not high then the benefit derived from the incentive systems may be outweighed by the resources expended implementing them. This is analogous to hiring security guards at a cost that is greater than the value of what they have been hired to guard.

2.3. Trust Management Systems

The other main form of inducing cooperation is trust management systems[2,4,7,16]. Generally, these systems work by having nodes within the network exchange reputation information. When a node detects uncooperative behaviour it disseminates this observation to other nodes which take action to avoid being affected by the node in question by changing traffic routes. In addition, some systems punish misbehaving nodes by isolating them from the network for a certain period of time in order to provide an incentive for users to cooperate. Note that although some trust management systems are also used to prevent malicious attacks, in this paper we are only concerned with the incentives aspects.

As with the token-based incentives system, trust management systems are subject to some significant problems. The first problem is that they take up considerable resources due to the constant transmission of observation data, which serves no purpose other than to monitor node behaviour. This hogs valuable processor clock cycles, memory, bandwidth and battery power that could be used to send actual data.

Trust management systems also suffer from vulnerabilities due to exchanging second hand information. Nodes may falsely accuse other nodes of misbehaving or collude with each other to cheat other users on the network. Although systems which rely only on first hand information have been investigated, they suffer from sensitivity to parameter settings as well as a lessened ability to punish uncooperative nodes[4]. They also do not take collusion of nodes into account.

Making decisions on whom to believe in a self-organising ad hoc network is very hard, requiring authentication as well as trust information about the accusing node. In practice this is extremely difficult to achieve, requiring either nodes which are known to be trustworthy (impractical for ad hoc networks) or bootstrapping trust relationships which involve significant complexity and risk, and may not be possible at all for dynamic or short-lived networks[4].

These factors make it questionable whether a trust management system could be effectively implemented in reality at a reasonable cost.

In addition, there have been very few experimental tests of either type of incentives systems to date. Almost all results come from simulations, which operate under assumptions and limited conditions that do not accurately reflect reality, and most importantly do not take user behaviour into account. Real life situations are invariably more complex and humans are often irrational and unpredictable, therefore, although the systems can be shown to work reasonably in simulations, real life implementations may show completely different results.

2.4. Transparency vs Choice

Incentives are by definition an inducement to stimulate or spur-on activity. In this case, we seek a method to induce users to cooperate with other users by allowing their devices to forward messages. Broadly speaking, this means that if given a choice, we want users to choose to allow forwarding the majority of the time, and to keep their devices on for forwarding even when they are not being used by the user.

It thus makes sense to consider how much choice a user should be given in the first place. We can choose to either have a system which is completely transparent and operates behind the scenes without the knowledge of users, or a system that users are aware of and can adjust themselves.

The less transparent the system is, the more complex it becomes for the user. At one extreme we might imagine a sending node having the option to choose between paths every time it sends information, with faster routes being more expensive and slower paths being cheaper. At the same time, every user of every intermediate node might have the option of choosing whether or not they wished to allow the hop and how much to charge for it. Considering how many times this process would need to repeated, if user intervention was needed each time this occurred it would be extremely inconvenient in practice.

A more reasonable middle ground would be to have agents which handled forwarding decisions according to preset rules, based on criteria such as the battery level and the current token store. However, given that the incentives system makes cooperation mandatory in order to forward, there would be little difference in the way that an agent made decisions compared with a human user, since they would both inevitably have to choose to forward most of the time and only stop when battery levels were low.

This then almost completely nullifies the whole point of having an incentives system since the user is essentially unaware of what is going on, and the agent behaviour (to forward the majority of the time and only stop or minimise forwarding when resources are scarce) might as well be hardwired into the system and work transparently behind the scenes. Users therefore do not need to be given any choice in the matter as it does not provide any additional utility to them and in fact may make devices less user friendly.

2.5. Proposed Solution

Given all the issues highlighted previously, it seems that ad hoc incentive systems as currently envisioned will not work successfully and ironically may cause more problems than they solve. In fact it is questionable whether incentive systems are necessary at all.

As stated in the introduction, user cooperation is only an issue when battery or other resources are scarce. Depending on the application, devices and users concerned, this may not even be an issue. As long as users are not unduly affected by forwarding for others, there should be little reason why they should not want to cooperate, especially if not cooperating requires more effort than cooperating.

In order for mobile ad hoc networks or indeed any new technology to move from concept to reality, it needs to go through successive phases of development, deployment and adoption in order to eventually achieve critical mass and enter the mainstream market. At each phase of technology adoption, there is a different target customer segment with different needs and preferences. Solutions should therefore be designed and implemented with each segment's unique needs in mind.

For ad hoc networks in particular, there is a need to work in distinct phases with the aim of steadily building up users. There is a chicken and egg situation where the usefulness of the network increases with the number of users forming and contributing to the network, but without enough users joining in initially, it will not be useful enough to attract more users. That is why a phased deployment makes much more sense than a full-scale deployment. Trying to run before being able to walk may result in the technology never taking off at all.

Unfortunately, current research into mobile ad hoc networks has mainly been conducted under the assumption that the networks will be mainly used for large-scale general consumer applications, and that nodes will be ubiquitous and reasonably dense. Both of these assumptions are considerably far from reality and will certainly not be true for initial phases of deployment; if the networks are designed and implemented with these assumptions in mind they run a high risk of failing. It is unreasonable to make plans for a bright future without first considering how to get there in the first place; the needs of the early market must not be ignored.

Given the strengths and weaknesses of ad hoc networks, it is unlikely that they will be able to be deployed on a large scale for general applications until much further down the adoption cycle. In the early stages, it is much more reasonable to expect ad hoc networks to be used for specific applications which fully capitalise on their strengths, with solutions that are both useful and financially sustainable[13]. In the same vein, it is unrealistic to expect a sudden proliferation of devices and networks having hundreds or thousands of nodes, especially with general applications that do not belong to a single authority.

In order to bootstrap adoption of the technology, it is therefore imperative that issues such as overly complex incentive systems do not cause early adopters of the technology to shun it. Early stage networks will most likely either be formed for specific applications under a single authority, where incentives are not needed, or by small groups of pioneering, technologically savvy users.

2.5.1. *Adoption Cycle For Mobile Ad Hoc Networks*

We therefore propose a solution that evolves according to the adoption cycle of mobile ad hoc networks, loosely based on Geoffrey Moore's Crossing the Chasm model[12]. In the earliest stage, we expect users to mainly be comprised of *pioneers*, technologically savvy users who are very enthusiastic about new technology and are more interested in exploring technology than actually benefiting from it. These users are very cooperative by nature and in addition are likely to be much more forgiving of faults in developing technologies; in many cases actually contributing to its development. We can draw parallels with the case of Peer-to-Peer networks, which usually see an extremely high level of cooperation in the early days, which declines slowly as they become more mainstream and attract more general users.

At this stage, we argue that incentive systems are not needed at all; the desired behaviour for nodes can simply be hardwired into nodes at a hardware as well as a protocol level and trust that the majority of users will not tamper with the devices. This will avoid all the problems discussed previously, ease implementation, reduce complexity and allow all forwarding functions to be handled automatically within the network for it to be fully self-organising.

Pioneering users have little incentive to hack the system and early applications are likely to be both specific and limited to small groups of users with common goals. By reducing problems and limitations for users, pioneers will become champions of the technology and introduce it to the next customer segment down the adoption cycle, the *visionaries*.

Visionaries are different from pioneers in that they are not interested in technology for technology's sake but rather see the potential in new technology and are willing to make sacrifices in order to be amongst the first to see that potential realised, and thereby get a head start in reaping the benefits. Visionaries are also likely to use the technology for specific applications, although the number of users may be significantly larger.

At this stage, incentive systems are again unnecessary as users of specific applications have implicit shared goals. There is also an inherent self interest for visionaries to see the technology that they choose succeed. Once there is a strong enough build up of visionaries and the technology has proven its worth, it is then possible to make the leap from the early market to the mainstream market, where the *pragmatists* await.

Pragmatists want a product that works and unlike the customers in the early market are much less tolerant of faults. They want to be able to buy products that meet their needs out of the box and easily get support from people who have used the technology before as well as find books about it in the bookstore. In short, they want a complete solution rather than a product that is still in development.

At this point of the technology's adoption, devices are reasonably ubiquitous and the technology has advanced beyond what was available in the early days. Most importantly, there are now a lot of experimental results and experience with real life implementations of the technology; it is also better understood how people actually use and abuse the system.

It is only at this point in the adoption cycle that it may make sense to introduce some form of incentives system. Even then, it would be better to design these incentives specifically for individual applications, based on what has been learned about how people abuse the networks, rather than a general incentives system that would possess the flaws discussed previously. As discussed in[13], it is unlikely that large-scale ad hoc networks will be deployed for general consumer applications due to their limitations in comparison to competing technologies. Their strengths will best be shown in either small-scale general applications or specific larger scale applications. In both cases, incentives can stem from common interest rather than an enforced system.

Finally, should mobile ad hoc networks become truly ubiquitous and used for general applications, *conservatives* will hop onto the bandwagon, simply because they have no choice. Conservatives want products that are cheap and simple; they buy products only after everyone they know already owns one.

2.5.2. *Do We Really Need Incentive Systems?*

It is of course still possible that in practice users will not wish to cooperate, even in the early stages of adoption. One of the main reasons behind the research of these systems is that the hardware and software of nodes can be tampered with and their behaviour modified by users in such a way that the devices do not cooperate with others, in order to save resources. Although it is generally recognised that most users do not have the required level of knowledge and skills to modify nodes, there is concern that criminal organisations will have the resources and interest to produce and sell modified nodes on a large scale.

Our position is that the majority of users will only cheat when it is clearly beneficial to them and relatively easy to do. In the case of mobile ad hoc devices, it is unclear that there is significant benefit to be had from going to the trouble to modify devices just to save resources such as battery power, memory and processor clock cycles. Battery power is probably the most limited resource, and even that may not prove to be an issue to most users, as long as the devices do not require constant charges. Devices might also be designed with docking capabilities such that when the devices are stationary their reliance on battery power is reduced as well as improving functionality. This will also encourage users to keep devices on to forward data for others even when not in use. Also, if devices become truly ubiquitous the power needed for forwarding will decrease anyway as the distance from one hop to another becomes minimal.

While there are certainly many criminal organisations with the ability to modify devices on a large scale, there is very little incentive for them to do so, since it is doubtful that a large enough market will exist to make the exercise profitable. A prominent example of a consumer device that has fallen victim to large scale tampering is the Sony Playstation 2 which has spawned an entire side industry of illegal modifications. Mod chips are widely available to buy on the Internet for home modification, as are full service organisations that modify units on behalf of consumers for a fee.

In the case of the Playstation, there are compelling reasons for both individual consumers and criminal organisations to engage in modification. Although the cost is relatively high, consumers who modify their devices can subsequently make significant savings by buying pirated games at a fraction of the original price. The organisations thus have a large and willing market of customers for their modifications, and are able to charge a significant sum to make large profits.

Conversely, in the case of mobile ad hoc devices, practically the only benefit to consumers would be longer battery lives. It is somewhat unlikely that they would go out of their way and pay a premium to modify their devices to this end, especially when it might cost the same to simply buy an extra battery with the added benefit of not voiding the device warranty or breaking the law. With little demand and potential for profitability, criminal organisations will not go to the trouble to reverse engineer and modify devices.

In any case, it would be necessary to produce a unique ad hoc device for each different type of application (e.g., a multiplayer ad hoc gaming device would be significantly different from an in-car ad hoc communications system). The need to reverse engineer each type of device as opposed to just one standard device would further increase costs and complexity for criminals and make it even less feasible for large scale modifications to occur.

Finally, any organisation with the knowledge to tamper with these devices would know (or soon learn) that there is no long-term value proposition to be gained from large-scale modifications. Unlike peer-to-peer file sharing networks such as Kazaa or Gnutella that can function reasonably well even with a large number of freeloaders[1], mobile ad hoc networks rely on cooperation for basic functionality. The more devices that do not cooperate by relaying messages, the worse the overall performance of the network will be, until finally the network is completely useless. Therefore, whilst a few isolated individuals might choose to modify their devices, a criminal organisation would gain no long-term benefit from doing so since they would rapidly destroy the network along with their own customers.

2.6. Conclusions

In this chapter we have looked at the problem of cooperation within mobile ad hoc networks and have analysed the drawbacks of currently proposed incentive systems. We then argued that there might not be a need for incentive systems at all, especially in the early stages of adoption, where excessive complexity can only hurt the technology's deployment. We looked at the needs of different customers segments for each stage within the projected technology adoption cycle and proposed that incentive systems not be used until ad hoc networks enter mainstream markets.

Even then, incentive systems should be tailored to the needs of each individual application rather than a general solution that may be too flawed or technically demanding to be implemented in reality. Punishments/incentives other than the denial of service to misbehaving nodes might be considered as an alternative. For example, within a file sharing application, users might be punished by limiting their query returns, rather than ostracising them from the network completely.

History is littered with examples of great technologies that never saw the light of day due to deployments that attempted to achieve too much too fast, with no way of successfully monetising the technology or to build up acceptance; all of which are dangers that ad hoc networks face. It is important to remember that mobile ad hoc networks are only one of a host of competing technologies, and in order to successfully make it to the mainstream market its worth over competing technologies needs to be clear and proven to consumers.

An important caveat to note is that the problem of providing incentives to selfish nodes is a somewhat separate issue from preventing malicious attacks on the network. In this chapter we have addressed the problem of nodes that wish to maximise their personal utility of the network, whereas malicious users may be less concerned with personal gain and simply wish to attack the network. Therefore, whilst we argue that incentive systems may not be necessary, it is still imperative that there are mechanisms to guard against malicious attacks in order to maintain the reliability of the network.

Future work could include experimental trials with two separate mobile ad hoc networks; one with an incentives system and the other without. Comparisons might then be made to confirm whether an incentive-less system would work as well or better than one with an incentives system in place. However, the experiment would need to be carefully designed such that users would behave in the same way as normal users along the adoption cycle, as experimental volunteers are likely to be cooperative by nature. Indeed, there are a number of as yet unsolved technical barriers to creating a cheat-proof incentive system for a fully decentralised, ad hoc network: some type of tamper proof token system is required; such a system would need a mechanism to prevent forged tokens or double spending of tokens; a decentralised witness system could work statistically to mitigate problems, but would itself consume scarce resources (none the least, battery power).

In conclusion, it is unlikely that there is a perfect solution to the ad hoc incentives problem. Implementations of technology are always limited in reality by cost, human behaviour, complexity and resources. Indeed, there is often only a least bad solution that provides the best cost benefit ratio rather than a best solution. It is more important at this point that mobile ad hoc networks be given the space to grow and develop rather than to choke it with complicated solutions to problems that may not even exist, causing users to shun the technology.

References

1. Adar, E. and Huberman, B. A., "Free riding on Gnutella" First Monday, 5(10), 2000, http://www.firstmonday.org.
2. Anderegg, L., Eidenbenz, S., "Ad hoc VCG: A Truthful and Cost-Efficient Routing Protocol for Mobile Ad hoc Networks with Selfish Agents" 2003, 245-259. (Proc. of the 9th annual international conference on Mobile computing and networking (Mobicom), 2003)
3. Anderson, R. and Kuhn M., "Tamper Resistance – A Cautionary Note", 1996, 1-11. (USENIX Workshop on Electronic Commerce, 1996), http://www.usenix.org
4. Bansal, S. and Baker, M., "Observation based cooperation enforcement in ad hoc networks" Stanford University Technical Report, 2003.
5. Boyd, C., "Metcalfe's Law" 2004, http://www.mgt.smsu.edu/mgt487/mgtissue/newstrat/metcalfe.htm
6. Zhong, S., Yang, R., Chen, J., "A Simple, Cheat-Proof, Credit-Based System for Mobile Ad-hoc Networks" March 2003, (Proc. of INFOCOM 2003), http://www.ieee-infocom.org.
7. Buchegger, S., Boudec, J.L., "Performance Analysis of the CONFIDANT Protocol: Cooperation Of Nodes - Fairness in Dynamic Ad-hoc Networks" 2002, 226-236. (Proc. of IEEE/ACM Symposium on Mobile Ad Hoc Networking and Computing, 2002)
8. Buttyan, L. and Hubaux, J.P., "Enforcing Service Availability in Mobile Ad-Hoc WANs" 2000, 87-96. (Proc. of IEEE/ACM Workshop on Mobile Ad Hoc Networking and Computing, 2000)
9. Buttyan, L. and Hubaux, J.P., "Nuglets: a Virtual Currency to Stimulate

Cooperation in Self Organised Mobile Ad Hoc Networks" Technical Report EPFL, January 2001.
10. Buttyan, L. and Hubaux, J.P., "Stimulating Cooperation in Self-organising Mobile Ad Hoc Networks" ACM/Kluwer Mobile, October 2003.
11. Crowcroft, J., Gibbens, R., Kelly, F., and Ostring, S., "Modelling incentives for collaboration in Mobile Ad Hoc Networks" March 2003, (Proc. of Modeling and Optimization in Mobile, Ad Hoc and Wireless Networks (WiOpt), 2003), http://www-sop.inria.fr
12. Moore, G.A., *Crossing the Chasm*, (HarperBusiness), 1991.
13. Huang, E., Östring, S., Crowcroft, J. & Wassell, I., "Developing Business Models for MobileMAN" (Technical Report, Cambridge University Engineering Library, CUED/FINFENG/TR.475, Jan 2004).
14. Jakobsson, M., Buttyan, L. and Hubaux, J.P., "A micro-payment scheme encouraging collaboration in multi-hop cellular networks" LNCS 2742-Springer, 2003, 15-33 (Proc. of Financial Crypto, 2003).
15. Lamparter, B., Paul, K. and Westhoff, D., "Charging Support for Ad Hoc Stub Networks" *Computer Communications*, **26**, (2003), 1504-1514.
16. Michiardi, P., Molva, R., "CORE: A Collaborative Repudiation Mechanism to enforce node cooperation in Mobile Ad hoc Networks" 2002, 107-121. (Sixth IFIP conference on communications and multimedia security (CMS 2002)).

CHAPTER 3

Fair and Efficient Allocation of Resources in the Internet

Ronaldo M. Salles

Dept. de Engenharia de Sistemas, Instituto Militar de Engenharia
22290-270, Rio de Janeiro, Brazil
Email: salles@ieee.org

Javier A. Barria

Dept of Electrical & Electronic Engineering, Imperial College London
London SW7 2BT, United Kingdom
E-mail: j.barria@imperial.ac.uk

This chapter proposes mechanisms to mediate the allocation of resources in networks shared by multiple types of applications as the Internet today. Social Choice Theory and Welfare Economics provide the foundations for the decision criteria used. Three different mechanisms are studied: Utility-Based Bandwidth Allocation (UBBA), Utility-Based Admission Control (UBAC) and Utility-Based Scheduling (UBS). Such mechanisms can be employed to sort out trade-offs in various resource allocation problems found in the Internet environment.

3.1. Introduction

This chapter proposes mechanisms to mediate the allocation of resources in networks shared by multiple types of applications. As IP technology turned out to be the common transport to all sorts of traffic, multi-application network environments become increasingly predominant in todays telecommunication enterprises. However, the diverse set of QoS requirements posed by applications makes the resource allocation problem difficult to be worked out, especially when degradation on performance becomes mandatory.

An adequate solution to the resource allocation problem can be qualified according to two important principles: *efficiency*, there should be no other allocation able to improve the performance of all applications, and *fairness*, the allocation should not deliberately penalise an individual or group in the name of efficiency. Social Choice Theory and Welfare Economics are applied to study such trade-off, quantify both principles, and formalise the resource allocation problem. From this study three different mechanisms are derived to enforce a solution for resource allocation problems in multi-application network environments, the mechanisms are based on utility functions information[a].

The first mechanism[1], Utility-Based Bandwidth Allocation (UBBA), provides bandwidth allocation for the integration of elastic (continuous utility) and inelastic (discontinuous utility) applications. Hence, this mechanism can be applied as a solution for the best-effort starvation problem (when strict-QoS applications dominate the use of network resources and starve best-effort traffic) found on the Internet today.

The second mechanism[2], Utility-Based Admission Control (UBAC), controls the admission of different application traffic to the links using associated utilities as functions of blocking probability. UBAC employs load estimation techniques and use asymptotic theory to simplify the problem of link partition. Blocking under UBAC is such that a fair and efficient use of resources is maintained.

Finally the last mechanism[3], Utility-Based Scheduling (UBS), uses utility functions expressed in terms of delays associated to each application in order to schedule packets on nodes. A measurement procedure computes such delays from the statistics involving packet arrivals and departures. From the utility functions, a decision procedure is employed so that packets are scheduled providing fair and efficient use of resources among all applications sharing the link.

3.2. Fairness, Efficiency and Utility Functions

Fairness is naturally associated to the concept of equality, while efficiency is a common engineering goal in the solution of a problem. For instance, consider a situation where goods are to be divided among identical agents, it is expected that a fair decision would be: "divide goods in equal parts". If this is done with minimum waste – e.g. if no goods are left, we may say that

---

[a]In Economics, utility is generally defined as a measure of the happiness or satisfaction gained from a good or service. In computer networks we adopt a more precise definition: *utility function* represents how each particular application (user) perceives quality according to the amount of allocated resources.

the solution is also efficient. However, such symmetrical distribution does not seem to hold as a fair and efficient solution for every situation, particularly the ones where agents have different characteristics and preferences. For these cases a more elaborated approach is necessary.

In the above-mentioned *division problem* it is assumed the multiplicity of agents (which are not necessarily identical), and the limited amount of resources to be divided. Each agent has his own preferences according to the benefit he extracts from the resources. Preferences are ranked through a function, agent's *utility function*.

The *division problem* is usually studied under the the framework of social choice theory and welfare economics[4]. In this chapter we are going to apply such theory and framework to provide insights and help us sort out resource allocation problems over the Internet.

First, it is important to elaborate the notion of utility functions applied to networks. According to Shenker[5], utility functions are defined for networking applications as functions that map a service delivered by the network into the performance of the application. For instance, if we take the bandwidth as service we may have in the network a given application that tolerates well variations on bandwidth, while other applications are very sensitive to such variations. Figure 3.1 illustrates different types of applications along with the respective utilities as functions of bandwidth.

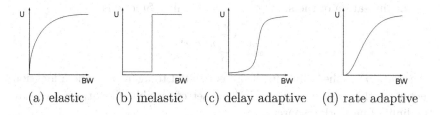

(a) elastic   (b) inelastic   (c) delay adaptive   (d) rate adaptive

Fig. 3.1. Example of utility functions associated to current Internet applications

Traditional Internet applications such as web, email and ftp are reported to be highly adaptive to the amount of bandwidth (elastic) being better described by strictly concave utility functions Fig. 3.1(a). Applications with hard real-time requirements (mission critic data, traditional telephony, high-quality video, CBR traffic) can be expressed by utilities on the form of Fig. 3.1(b), while delay and rate adaptive utilities can be associated to some multimedia applications – audio and video with "Internet quality".

Other examples of utility functions applied to networks are widespread in the literature and will be also presented here in the following sections. However to formalise the idea we present the definition used in this chapter which was extracted from the work of Sairamesh et al.[6].

Definition 3.1: Utility Functions for Computer Network Applications. Let $x = \{x_1, x_2, \ldots, x_M\}$ be the vector of allocated network resources. From that vector the corresponding set of QoS parameters is computed: $q = \{q_1, q_2, \ldots, q_n\}$, where each q_i is a function of $\Re^M \to \Re$. The utility function is defined by: $u(q)$, $\Re^n \to \mathcal{D} \subseteq \Re$.

The only requirements we impose on the utility functions, $u(q)$, is monotonicity.

Associating the most adequate utility function for each application we are ready to employ the theory of welfare economics to support the solution of resource allocation problems in multi-application network environments as the Internet.

As already mentioned the solution for a resource allocation problem should be *efficient*. In social choice theory and welfare economics *efficiency* is usually associated to the concept of *Pareto Efficiency*[7]: "an allocation is Pareto Efficient if no one can improve his benefit without decreasing the benefit of another". A solution based on this criterion can be achieved from the maximisation of the sum[8] of all agent utility functions,

$$\max_{x \in S} \sum_{i=1}^{N} u(x, i) \qquad (1)$$

where $u(x, i)$ is the utility function associated to agent i, x the allocation vector (in this case $q = x$), and S the set of feasible allocations given by the limited network resources.

The main concern, however, is that there is no fairness component embedded in the solution since the *Pareto Criterion* takes no interest in distributional issues no matter how unequally the solution may arise[9]. For instance, an allocation where agent 'A' gets all resources while agent 'B' gets nothing may still be *Pareto Efficient* (may still solve (1)). Moreover, there may be several *Pareto* solutions or even infinite depending on the type of utility functions and problem considered.

Fairness theory provides other better suited criteria to sort out the allocations specially when agents are different (different applications). One of the most widely accepted concept of fairness is due to Rawls[10]: "the system is no better-off than its worse-off individual", i.e. it implies the maximisation of the benefit (utility) of the worse-off agent,

$$\max_{x \in S} \min_i \{u(x,i)\} \tag{2}$$

This criterion usually leads to the equalisation of utility functions, in this case it is also known as the *egalitarian principle*. Note that utility equalisation has already been applied to network bandwidth allocation problems[11].

The problem on choosing one of the criterion above to sort out resource allocation problems is that with (2) *efficiency* can not be guaranteed, while with (1) *fairness* is often not achieved. This established the well-known trade-off[9] between *fairness* and *efficiency*. One alternative is to consider the lexicographic extension[12] of the Rawlsian criterion since it enjoys both *fairness* and *efficiency* properties discussed. A formal definition for the lexicographic allocation is presented below.

Definition 3.2: A feasible allocation $x : (x_1, x_2, \ldots, x_J)$ is lexicographically optimal if and only if an increase of any utility within the domain of feasible allocations must be at the cost of a decrease of some already smaller utility. Formally, for any other feasible allocation y, if $u(y,j) > u(x,j)$ then there must exist some k such that $u(x,k) \leq u(x,j)$ and $u(y,k) < u(x,k)$.

It is important to note that when the egalitarian criterion produces a solution where all the available resources are allocated, this solution also satisfies Def. 3.2 and so for this case egalitarian and lexicographic criteria are equivalent. We also explore this condition in the mechanisms described in the following sections.

3.3. Utility-Based Bandwidth Allocation

In this section we study the problem of bandwidth allocation in a network environment such as the Internet where elastic and inelastic applications coexist. We consider the utility functions associated to the applications as illustrated in Fig. 3.1(a) and Fig. 3.1(b) respectively.

It is important to note that while elastic applications may tolerate changes in performance and adapt to current network state, inelastic applications are non-adaptive in the sense that if the network is not able to provide the requested amount of bandwidth, utility will be zero. This behaviour is represented by the step functions in Fig. 3.1(b). Such functions do not facilitate the allocation problem and the integration with elastic applications. For instance, there is no way to apply the egalitarian principle unless all utilities are 1 (100% of performance). Utility aggregation techniques studied next overcome this difficulty and also guarantee scalability to the approach.

3.3.1. Utility of the Aggregate

Applications may have their individual traffic flows aggregated into a single flow to reduce the amount of information used by management procedures and to simplify network mechanisms over the Internet. In this case, to apply the allocation criterion an utility function should be also determined for the aggregated flow.

For elastic applications it can be seen that the same type of utility function can be used for the aggregate. Let us assume that n adaptive flows are using the end-to-end path p, let x_i be the allocation of flow i and $u(x_i, i)$ its utility function. By the egalitarian criterion we must have $u(x_i, i) = u(x_j, j)$ for any other flow j, and thus $x_i = x_j = x$ since their utility functions are identical. Hence, the aggregate flow on path p is allocated $X_p = nx$ and can be associated to an utility function $u(X_p, p)$ which is identical to any $u(x_j, j)$ with the x-axis multiplied by n.

The procedure just described does not serve for the case of inelastic applications since it will also produce a step utility function for the aggregate. The utility of an aggregate of inelastic flows does not seem to be described by such binary relations. Given their strict requirements such flows are subjected to admission control procedures in order to guarantee the requested QoS (step utility function). The relationship between flows that go through and flows that are blocked provides the key issue to establish the aggregate utility. In other words, the amount of aggregate traffic effectively transmitted through the network should indicate the utility experienced by the aggregate as a whole according to network services.

From the aggregate viewpoint the utility should be expressed by the relation between the *aggregate carried traffic* (C.T.) and the *aggregate offered traffic* (O.T.) so that if all offered traffic is accepted and resources allocated to conform with individual requirements, the aggregate may have reached 100% of satisfaction. In this sense, different levels of satisfaction (utility) can be associated to other percentages. Figure 3.2 illustrates such function where r is given by the ratio C.T./O.T. In this particular case the aggregate level of satisfaction (utility) increases in the same proportion (linear) as the network is able to carry more of its traffic. Also, when just 50% or less of its offered traffic is carried there will be no utility for the aggregate.

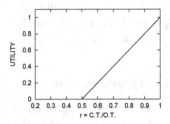

Fig. 3.2. Utility function for the aggregate of inelastic flows.

Note that the parameter $1 - r$ represents the *blocking probability* and can be used to relate the aggregate utility with the allotted bandwidth. Assume that each aggregate s is allocated a certain amount of bandwidth X_s in all the links along its path so that links reserve the required resources based on a *complete partitioning* approach. This system can be described by the *Erlang Model*[13] with blocking probabilities given by

$$1 - r_s = \frac{\frac{\rho_s^{N_s}}{N_s!}}{\sum_{i=0}^{N_s} \frac{\rho_s^i}{i!}} \qquad (3)$$

where ρ_s represents the *offered traffic* (O.T.), and $N_s = \lfloor X_s/b_s \rfloor$ the maximum number of individual flows that aggregate s may have for the reserved X_s. Once ρ_s is determined, the utility function for the aggregate can be expressed as a composite function of X_s:

$$\text{aggregate utility: } u(r_s(X_s, \rho_s), s) \qquad (4)$$

Note that this function conforms with Def. 3.1.

From the egalitarian principle the bandwidth allocation solution for a system where an aggregate j of adaptive flows and an aggregate s of non-adaptive flows share the same path, is given by:

$$u(X_j, j) = u(r_s(X_s, \rho_s), s) \qquad (5)$$

Thus, X_j and X_s must satisfy

$$\frac{\frac{\rho_s^{\lfloor X_s/b_s \rfloor}}{\lfloor X_s/b_s \rfloor!}}{\sum_{i=0}^{\lfloor X_s/b_s \rfloor} \frac{\rho_s^i}{i!}} + u^{-1}(u(X_j, j), s) = 1 \qquad (6)$$

$$X_j + X_s = C \qquad (7)$$

where C is the minimum capacity among the links along the path. Given ρ_s and b_s, it is difficult to find and also unlikely to exist X_s and X_j satisfying (6)–(7). Although we managed to derive utility functions for the aggregates of non-adaptive flows that could be expressed in terms of the allotted bandwidth X_s, the composite relation in (4) make those utility functions not directly applicable to our allocation scheme.

3.3.2. *Limiting Regime Approximation*

It can be shown that expressions can be strongly simplified if the network is assumed to be in the *limiting regime*, i.e. when $N_s = \lfloor X_s/b_s \rfloor$ is large and $\rho_s > N_s$ (these two conditions are likely to be verified in today's backbone links). In this case the following approximation is valid (see proof in the Appendix):

$$r \sim \frac{X}{b\rho} \qquad (8)$$

With the approximation above,

$$u(r_s(X_s, \rho_s), s) \sim u\left(\frac{X_s}{b_s \rho_s}, s\right) \qquad (9)$$

Therefore we eliminate the need for the composite relation and the utility of the aggregate can be directly obtained as a function of bandwidth without using the intermediate equation in (3). Equation (8) provides a straightforward way to relate the parameter r with the bandwidth allotted to the aggregate. From the egalitarian criterion, X_j and X_s are now obtained from:

$$u(X_j, j) = u(X_s/b_s\rho_s, s) \qquad (10)$$
$$X_j + X_s = C \qquad (11)$$

which can be solved for continuous and monotone utility functions employing simple computational methods (e.g. Newton's method, non-linear programing techniques[14], piecewise linear approximation[11]).

The solution of (10) provides a *fair* (in the Rawlsian sense) allocation of network resources between elastic and inelastic flows. Moreover, if all resources are allotted the solution is also *efficient*. Note that our approach works at the aggregate level and so it is only necessary to use information from aggregate utility functions. The number of adaptive flows n and offered traffic ρ should be determined in advance in order to obtain aggregate utilities for adaptive and non-adaptive flows respectively. After new estimates are obtained, (10) should be solved again to generate updated allocations. Finally those allocations can be implemented in network links using scheduling mechanisms, for instance weighted fair queuing with weights $w_j = X_j/C$ and $w_s = X_s/C$. The next subsection proposes a way to estimate the offered load ρ

3.3.3. Offered Load Estimation

In order to obtain the aggregated utility functions for non-adaptive applications as in (9) it is necessary to first compute the *offered load* ρ of each aggregated flow. Since aggregated loads are not known in advance and may vary from time to time, it is necessary to update them so as to provide the correct information to the allocation schemes. We present in this Subsection a technique based on the *maximum likelihood estimation* (MLE) that continuously estimate aggregate loads from network observed states.

The complete network state at time t_i is given by the number of individual connections being served inside the aggregated flows, $\boldsymbol{n}(i)$: $(n_1(i), n_2(i), \ldots, n_k(i))$, where each $n_s(i)$ indicates the observation at time t_i for aggregated flow s. Let us focus on a single aggregated flow and its state $n(i)$. We assume the state sampled at each time window, t, and the estimator updated using $R = T/t$ samples as depicted in Fig. 3.3.

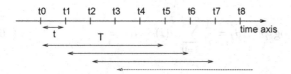

Fig. 3.3. Time windows used for load estimation

We assume that we have at each time a set of R independent samples which can be used to estimate the aggregate loads. The aggregate of non-adaptive flows is reserved a portion of link resources along all its path, so that each link operates under a *complete partitioning* (CP) policy. The CP policy is equivalent to a set of independent Erlang systems[13], therefore the following *likelihood function* is applied to the aggregate samples,

$$p(n(i); \theta) = \frac{\theta^{n(i)}}{n(i)!} G^{-1}(\theta); \quad G(\theta) = \sum_{j=0}^{C} \frac{\theta^j}{j!} \quad (12)$$

Taking all the R samples, the *likelihood and log-likelihood functions* are given by:

$$p(\boldsymbol{n}; \theta) = \frac{\theta^{[\sum_{i=0}^{R-1} x(i)]}}{\prod_{i=0}^{R-1} n(i)!} G^{-R}(\theta) \quad (13)$$

$$\log p(\boldsymbol{n}; \theta) = \log \theta \sum_{i=0}^{R-1} n(i) - \sum_{i=0}^{R-1} \sum_{j=0}^{n(i)-1} \log(n(i) - j) - R \log G(\theta) \quad (14)$$

The MLE is computed from

$$\frac{\partial}{\partial \theta} \log p(\boldsymbol{n}; \theta) = \frac{1}{\theta} \sum_{i=0}^{R-1} n(i) - R \frac{\sum_{j=1}^{C} \frac{\theta^{j-1}}{(j-1)!}}{\sum_{j=0}^{C} \frac{\theta^j}{j!}} \quad (15)$$

The fraction in the right hand side term is simply the probability of connection acceptance: r. Thus the algorithm is reduced to

$$\frac{\partial}{\partial \theta} \log p(\boldsymbol{n}; \theta) = \frac{1}{\theta} \sum_{i=0}^{R-1} n(i) - Rr = 0 \quad (16)$$

$$\hat{\theta} = \frac{\sum_{i=0}^{R-1} n(i)}{R(1 - \hat{q})} \quad (17)$$

The parameter $\hat{q}$ can be directly computed at each time window T through the ratio between the number of blocked connections and the total number of connection arrivals. In fact the input data should also include this measure: $[n(0), n(1), ..., n(R-1), \hat{q}]$. Therefore, $\hat{\theta}$ provides an estimate for the missing parameter in the utility of the aggregate of inelastic flows (9).

3.4. Utility-Based Admission Control

In this section utility function information is used to support admission control decision for different types of inelastic application flows that enters the network. Based on such information, admission control decisions provide a *fair* and *efficient* use of available resources.

One important performance parameter for the case of admission control is *blocking probability*, utility functions should be expressed in terms of this parameter. Similarly to what has been done in Sec. 3.3, for an aggregate of inelastic flows s the utility function is given by $u(q_s, s)$, where q_s is the *blocking probability* associated to flows. Note that this type of utility function conforms with Def. 3.1.

To simplify the analysis and use the formulation introduced in the previous section we assume utilities are linear on q_s, $u(q_s, s) = 1 - \delta_s . q_s$, where the weight δ_s indicates the sensitivity of the aggregate in relation to *blocking probability*. The lower the *blocking probability*, the higher the utility. When $q_s = 0$ we have $u = 1$.

The weight δ_s may also serve to give preferences for some aggregate traffic over others on the access to the network. From the *limiting regime approximation* presented in Subsection 3.3.2 we have:

$$u(q_s, s) = 1 - \delta_s \left(1 - \frac{X_s}{b_s \rho_s}\right) \quad (18)$$

Let us assume that K aggregates access a network link l with available bandwidth C, using utility equalisation it follows that,

$$u(r_1, 1) = u(r_2, 2) = \ldots = u(r_K, K) \quad (19)$$

$$\delta_1 \left(1 - \frac{X_1}{b_1 \rho_1}\right) = \delta_2 \left(1 - \frac{X_2}{b_2 \rho_2}\right) = \ldots = \delta_K \left(1 - \frac{X_K}{b_K \rho_K}\right) \quad (20)$$

$$\sum_{i=1}^{K} X_i \leq C \quad (21)$$

If condition (20) is satisfied as an equality, the allocation returned $\{X_1, \ldots, X_K\}$ by utility equalisation is also lexicographically optimal (see Def. 3.2).

However, there is a fundamental issues that makes the approach above not straightforward. It is not always possible for any set of weights $\{\delta_1, \ldots, \delta_K\}$ to have the equality on (19) being satisfied. For instance, taking the aggregate of minimum weight ($\delta_j \leq \delta_k$) as reference the partitions that solve (19) are given by:

$$X_k = b_k\rho_k\left[1 - \frac{\delta_j}{b_j\rho_j\delta_k}(b_j\rho_j - X_j)\right], \quad k \neq j \tag{22}$$

$$\sum_{k=1}^{K} X_k \leq C \tag{23}$$

In order to guarantee $X_j > 0 \Rightarrow$

$$X_k > b_k\rho_k\left(1 - \frac{\delta_j}{\delta_k}\right) \Rightarrow C > \sum_{k=1}^{K} b_k\rho_k\left(1 - \frac{\delta_j}{\delta_k}\right) \tag{24}$$

Given the reserved link bandwidth C, it is not always possible to solve (19) for general values of b_k, ρ_k and δ_k. It is necessary to apply another criterion to relax the equality condition on (19). Hence, we build up the Utility-Based Admission Control (UBAC) service through the Rawlsian criterion so that *fairness* is achieved although *efficiency* is not guaranteed.

Recalling the Rawlsian criterion defined in Sec. 3.2, link partitions in the UBAC are then obtained from the problem:

$$\text{find: } (X_1, X_2, \ldots, X_K) \quad N_k = \lfloor X_k/b_k \rfloor \tag{25}$$

$$\text{s.t.: } \max \min_k \{u(q_k, k)\} \quad k = 1, 2, \ldots, K \tag{26}$$

$$\sum_{k=1}^{K} X_k \leq C \tag{27}$$

Substituting (18) in (26) we have,

$$\max \min_k \{1 - \delta_k.q_k\} \quad k = 1, 2, \ldots, K \tag{28}$$

which can be transformed to,

$$\min \max_k \{\delta_k.q_k\} \quad k = 1, 2, \ldots, K \tag{29}$$

Finally, with the introduction of an extra variable (ξ), the minmax in (29) can also be written as:

$$\xi^* = \min \{\xi \mid \delta_k q_k \leq \xi, \quad k = 1, 2, \ldots, K\} \tag{30}$$

Since there is no restriction on ξ apart from being a real number, it is always possible to return a solution for (30). The *blocking probability* for aggregate k is given by,

$$q_k \leq \frac{\xi^*}{\delta_k} \tag{31}$$

Substituting q_k in (30) by the corresponding expression given by the *limiting regime* approximation, we end up with the following linear programming problem,

$$\min \left\{ \xi \,\middle|\, \delta_k X_k + \xi a_k \geq \delta_k a_k,\ X_k < a_k,\ \sum_{k=1}^{K} X_k \leq C \right\} \quad (32)$$

where: $k = \{1, ..., K\}$, $a_k = b_k \rho_k$, and $\xi \in \Re$. The above problem has $K+1$ variables (all X_k's plus ξ) and $2K+1$ constraints, demanding low computational time to find a solution (K is the number of aggregates crossing the link which is assumed to be small). Thus, the UBAC is implemented through the partitions computed from (32). The partitions should be re-computed (problem (32) solved again) whenever the offered traffic load ρ_k of the aggregates changes.

3.5. Utility-Based Scheduling

In this section *Utility-Based Scheduling (UBS)* disciplines are proposed to support different classes of adaptive applications over the Internet. The disciplines differ from other well known techniques (e.g. WFQ, WRR) since scheduling decisions are directly controlled by utility function information. In this case utility functions describe the expected level of performance of each class, and rely on the acquisition of on-line traffic statistics to trigger scheduling decisions.

Although the formulation on Sec. 3.3 and Sec. 3.4 were similar, the approach here is different since UBS deals with packet level procedures. Given that the services in this section are expressed in terms of class (aggregate) packet delays, it is natural to associate to the classes utility functions that depend on those delays. Hence, in the point of view of each class the resource that is being allocated represents the "promptness" of the scheduler to deliver its packets.

There should be a maximum delay that each class tolerates, from that point there will be no utility for the class. For all other situations the class as an entity may associate a level of satisfaction corresponding to the promptness offered by the scheduler. Promptness can be measured by the expression,

$$P_k = \frac{D_k^{\max} - D_k}{D_k^{\max}} \quad (33)$$

where D_k is the current delay and $D_k^{\max}$ the maximum delay class-k tolerates.

When $D_k = 0$, $P_k = 1$ (100%) which means that the scheduler is always available to serve class-k packets. On the other hand, when $D_k = D_k^{\max}$ class-k is completely starved resulting in $P_k = 0$. In between these two extreme scenarios different levels of performance are achieved and can be associated to a general utility function that depends on P_k: $u(P_k, k)$. To be consistent the utility should be monotonically increasing on P_k, or in the other way around monotonically decreasing in D_k. Again, such utility function conforms with Def. 3.1.

The goal defined for the scheduler may be to guarantee utility equalisation. The only action the scheduler may perform to equalise those utilities is to select at each departure epoch an adequate class to be served. Given packet dynamics and the limited control of the scheduler over D_k, such equalisation at any time instant t is virtually impossible. The best the scheduler can do is to select for service at each departure epoch the class of minimum utility in an attempt to increase its utility and move towards equalisation. The UBS is defined in this way as seen in (34) for general utility functions $u(P_k, k)$.

$$\max_t \min_k \{u(P_k, k)\} \quad k-1\ldots K \tag{34}$$

Note that the objective of the UBS is to maximise the minimum utility in the system and so it conforms with the *fairness* concept introduced (Rawlsian criterion). Note also that the UBS may emulate the Proportional Differentiated Services (PDS) proposed by Dovrolis[15] if utility functions are such that:

$$u_k = \delta_k D_k^{\max}(1 - P_k) \tag{35}$$

In this way, the scheduler will be always enforcing the PDS equality $\delta_1.D_1 = \delta_2.D_2, \ldots, \delta_K.D_K$, where δ_k is the weight associated to class-k and D_k the respective delay.

3.5.1. *Measuring Class Delays*

The utilities defined for the UBS use class packet delays D_k as argument for the functions, therefore it is necessary to derive a procedure to obtain such delays.

Each parameter D_k is defined in the UBS as the average time a packet from class-k has to wait on a queue before starting transmission. If this measure is computed only for long-term scales, applications that lasts a shorter period of time may not benefit from UBS services. We propose a measure scheme for the class delays that takes into account current and past packet queueing delays. The approach can be configured to the time scale of interest and provides low overhead on computations.

For each class in the system we define a measurement window to store information about most recent arrival/departure statistics. The window is implemented by a circular list with two pointers: *head-of-window* (hw) and *head-of-queue* (hq). Figure 3.4 illustrates the mechanism of the windows.

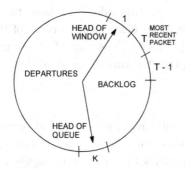

Fig. 3.4. Measurement window implemented by a circular list.

Both pointers move in the counter clockwise direction as packets enter and leave the system, when the system is empty the pointers are aligned. An incoming packet overwrites the hw register with its new information and shifts the pointer to the next position. The hq points to the oldest packet in the queue and when this packet is served it is also shifted. The information stored consists of time-stamps of packets in queue, and waiting times of packets that have already been served. For each class-k we have an associated window T_k of size $|T_k|$ and the following notation,

Thus, the mean delay on window k is a function of time and can be computed from:

$$D_k(t) = \frac{1}{|T_k|} \sum_{i \in T_k} d_k^i = \frac{1}{|T_k|} \left(\sum_{i \in B_k} d_k^i + \sum_{i \notin B_k} d_k^i \right)$$

$$= \frac{1}{|T_k|} \left(n_k t - \sum_{i \in B_k} \tau_k^i + \sum_{i \notin B_k} d_k^i \right)$$

$$= \frac{1}{|T_k|} (n_k t - S_k + D_k^{dep}) \qquad (36)$$

where the variables n_k, S_k, D_k^{dep} are updated as indicated next.

At each arrival epoch:
$n_k \leftarrow n_k + 1$
$S_k \leftarrow S_k + \text{clock}$
$D_k^{dep} \leftarrow D_k^{dep} - d_k^{hw}$

At each service epoch:
$n_k \leftarrow n_k - 1$
$S_k \leftarrow S_k - \tau_k^{hq}$
$d_k^{hq} \leftarrow \text{clock} - \tau_k^{hq}$
$D_k^{dep} \leftarrow D_k^{dep} + d_k^{hq}$

Note that the measure procedure does not need to set or modify any packet field, and that the mean delay as a function of time is readily obtained from simple computations. The size of the windows $|T_k|$ can be changed as desired to capture the time-scale of interest for the service.

3.6. Conclusion

This chapter studied the problem of resource allocation in multi-application network environments such as the Internet. We applied the theory of Social Choice and Welfare Economics to derive some useful criteria that could be applied to our problem. Two important properties, *fairness* and *efficiency*, were defined and used to qualify the solution able to mediate inherent trade-offs found in resource allocation problems. Three different mechanisms were studied: bandwidth allocation (UBBA), admission control (UBAC) and packet scheduling (UBS). Such mechanisms provide invaluable solutions for current resource allocation problems and can be directly implemented in the network. Simulation results evaluating the performance of all mechanisms can be found in the PhD thesis[16] that originated this work. Future work will be done in the practical implementation of the approach in real network scenarios.

Acknowledgments

This work was supported in part by CNPq Brazil, under Grant: 200049/99.

Appendix - Limiting Regime Approximation

Let $N = \lfloor X/b \rfloor$, the *Limiting Regime* is achieved when $N \to \infty$ with $\rho > N$, which represents congested network backbone links. From Jagerman[17] we have the following *Erlang B* equivalent formula,

$$B(N,a)^{-1} = \int_0^\infty e^{-t} \left(1 + \frac{t}{a}\right)^N dt \qquad (37)$$

Let $\alpha = \frac{N}{\rho} < 1$,

$$B(N,\rho)^{-1} = \int_0^\infty e^{-t}\left(1 + \frac{\alpha.t}{N}\right)^N dt \qquad (38)$$

in the limit as N grows,

$$\lim_{N\to\infty} B(N, \frac{N}{\alpha})^{-1} = \int_0^\infty e^{-t}. \lim_{N\to\infty}\left(1 + \frac{\alpha.t}{N}\right)^N dt \qquad (39)$$

$$= \int_0^\infty e^{-t}.e^{\alpha.t} dt = \int_0^\infty e^{(\alpha-1)t} dt \qquad (40)$$

Since $(\alpha - 1) < 0$ we have,

$$\lim_{N\to\infty} B(N, \frac{N}{\alpha}) = -(\alpha - 1) = 1 - \alpha \qquad (41)$$

Thus, *blocking probabilities* converges to $1 - \frac{X}{b\rho}$, which completes the proof. □

References

1. Salles, R.M. and Barria, J.A.,"Fair bandwidth allocation for the integration of adaptive and non-adaptive applications" LNCS 3126- Springer, 2004, 1-12. (Proc. of the 8th IEEE Inter. Conf. on Telecomm.,2004)
2. Salles, R.M. and Barria, J.A.,"Proportional differentiated admission control", *IEEE Comm. Let.*, **8**(5), (2004) 320 -322.
3. Salles, R.M. and Barria, J.A., "Utility-based scheduling disciplines for adaptive applications over the internet",*IEEE Comm. Let.*, **6**(5), (2002) 217 -219.
4. Sen, A.K., *Collective Choice and Social Welfare*, (Ed. North-Holland Publishing) 1995.
5. Shenker, S., "Fundamental design issues for the future internet" *IEEE JSAC*, **13**(7), (1995) 1176-1188.
6. Sairamesh, J., Ferguson, D.F. and Yemini, Y.,"An approach to pricing, optimal allocation and quality of service provisioning in high-speed packet networks ", 1995, 1111-1119, (Proc. of the IEEE INFOCOM'95, 1995).
7. Mas-Colell, A., Whinston, M.D. and Green, J.R., *Microeconomic Theory*, (Oxford University Press), 1995.
8. Kelly, F.P., Maulloo, A.K. and Tan, D.K.H., "rate control for communication networks: shadow prices proportional fairness and stability" *J. of the Oper. Res. Soc.*, **49**, (1998) 237–252.
9. Sen, A., "The possibility of social choice" *The Amer. Econ. Rev.*, **89**(3), (1999) 349-378.
10. Rawls, J., *A Theory of Justice*,(Oxford University Press), 1999.
11. Bianchi, G. and Campbell, A.T.,"A programmable MAC framework fir utility-based adaptive quality of service support" *IEEE JSAC*, **18**(2),(2000) 244-255.

12. Roberts, K.W.S., "Interpersonal comparability and social choice theory" *Review of Economic Studies*, **47**(2), (1980) 421-439.
13. Ross, K.W., *Multiservice Loss Models for Broadband Telecommunication Networks*, (Springer-Verlag), 1995.
14. Fletcher, R., *Practical Methods of Optimization*, (John Wiley & Sons), 2000.
15. Dovrolis, C. and Ramanathan, P.,"A case for relative differentiated services and the proportional differentiation model " *IEEE Network Magazine*, **13**(5),(1999) 26-34.
16. Salles, R.M., *Fair and Efficient Resource Allocation Strategies in Multi-Application Networks using Utility Functions*, (Ph.D. Thesis, DEEE, Imperial College London), 2004.
17. Jagerman, D.L.,"Some Properties of the Erlang Loss Function" *The Bell Sys. Tech. J.*, **53** (1974) 525-550.

CHAPTER 4

The Locality Principle

Peter J. Denning

Naval Postgraduate School
Monterey, CA 93943 USA
Email: pjd@nps.edu

Locality is among the oldest systems principles in computer science. It was discovered in 1967 during efforts to make early virtual memory systems work well. It is a package of three ideas: (1) computational processes pass through a sequence of locality sets and reference only within them, (2) the locality sets can be inferred by applying a distance function to a program's address trace observed during a backward window, and (3) memory management is optimal when it guarantees each program that its locality sets will be present in high-speed memory. Working set memory management was the first exploitation of this principle; it prevented thrashing while maintaining near optimal system throughput, and eventually it enabled virtual memory systems to be reliable, dependable, and transparent. Many researchers and system designers rallied around the effort to understand locality and achieve this outcome. The principle expanded well beyond virtual memory systems. Today it addresses computations that adapt to the neighbourhoods in which users are situated, ways to infer those neighbourhoods by observing user actions, and optimising performance for users by being aware of their neighbourhoods. It has influenced the design of caches of all sorts, Internet edge servers, spam blocking, search engines, e-commerce systems, email systems, forensics, and context-aware software. It remains a rich source of inspirations for contemporary research in architecture, caching, Bayesian inference, forensics, web-based business processes, context-aware software, and network science.

4.1. Introduction

Locality of reference is one of the cornerstones of computer science. It was born from efforts to make virtual memory systems work well. Virtual memory was first developed in 1959 on the Atlas system at the University of Manchester. Its superior programming environment doubled or tripled programmer productivity. But it was finicky, its performance sensitive to the choice of replacement algorithm and to the ways compilers grouped code on to pages. Worse, when it was coupled with multiprogramming, it was prone to thrashing, the near-complete collapse of system throughput due to heavy paging. The locality principle guided us in designing robust replacement algorithms, compiler code generators, and thrashing-proof systems. It transformed virtual memory from an unpredictable to a robust technology that regulated itself dynamically and optimised throughput without user intervention. Virtual memory became such an engineering triumph that it faded into the background of every operating system, where it performs so well at managing memory with multithreading and multitasking that no one notices.

The locality principle found application well beyond virtual memory. Today it directly influences the design of processor caches, disk controller caches, storage hierarchies, network interfaces, database systems, graphics display systems, human-computer interfaces, individual application programs, search engines, Web browsers, edge caches for Web based environments, and computer forensics. Tomorrow it may help us overcome our problems with brittle, unforgiving, unreliable, and unfriendly software.

I will tell the story of this principle, starting with its discovery to solve a multimillion-dollar performance problem, through its evolution as an idea, to its widespread adoption today. My telling is highly personal because locality, and the attending success of virtual memory, was my focus during the first part of my career.

4.2. Manifestation of a Need (1949-1965)

In 1949 the builders of the Atlas computer system at University of Manchester recognised that computing systems would always have storage hierarchies consisting of at least main memory (RAM) and secondary memory (disk, drum). To simplify management of these hierarchies, they introduced the page as the unit of storage and transfer. Even with this simplification, programmers spent well over half their time planning and programming page transfers, then called overlays. In a move to enable programming productivity to at least double, the Atlas system builders therefore decided to automate the overlaying process. Their "one-level storage system" (later called virtual memory) was part of the second-generation Atlas operating system in 1959[30]. It simulated a large main memory within a small real one. The heart of their innovation was the novel concept that addresses named values, not memory locations. The CPU's addressing hardware translated CPU addresses into memory locations via an updatable page table map (Fig. 4.1). By allowing more addresses than locations, their scheme enabled programmers to put all their instructions and data into a single address space. The file containing the address space was on the disk; the operating system copied pages on demand (at page faults) from that file to main memory. When main memory was full, the operating system selected a main memory page to be replaced at the next page fault.

The Atlas system designers had to resolve two performance problems, either one of which could sink the system: translating addresses to locations; and replacing loaded pages. They quickly found a workable solution to the translation problem by storing copies of the most recently used page table entries in a small high speed associative memory, later known as the address cache or the translation lookaside buffer. The replacement problem was a much more difficult conundrum.

Because the disk access time was about 10,000 times slower than the CPU instruction cycle, each page fault added a significant delay to a job's completion time. Therefore, minimising page faults was critical to system performance. Since minimum faults means maximum inter-fault intervals, the ideal page to replace from main memory is the one that will not be used again for the longest time. To accomplish this, the Atlas system contained a "learning algorithm" that hypothesised a loop cycle for each page, measured each page's period, and estimated which page was not needed for the longest time.

The learning algorithm was controversial. It performed well on programs with well-defined loops and poorly on many other programs. The controversy spawned numerous experimental studies well into the 1960s that sought to determine what replacement rules might work best over the widest possible range of programs. Their results were often contradictory. Eventually it became apparent that the volatility resulted from variations in compiling methods: the way in which a compiler grouped code blocks onto pages strongly affected the program's performance under a given replacement strategy.

Meanwhile, in the early 1960s, the major computer makers were drawn to multiprogrammed virtual memory because of its superior programming environment. RCA, General Electric, Burroughs, and Univac all included virtual memory in their operating systems. Because a bad replacement

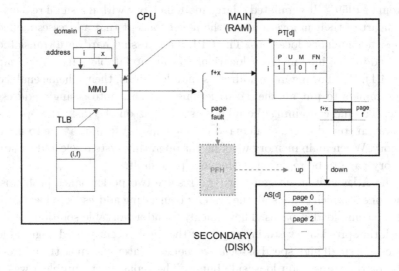

Fig. 4.1. The architecture of virtual memory. The process running on the CPU has access to an address space identified by a domain number d. A full copy of the address space is stored on disk as the file AS[d]; only a subset is actually loaded into the main memory. The page table PT[d] has an entry for every page of domain d. The entry for a particular page (i) contains a presence bit P indicating whether the page is in main memory or not, a usage bit U indicating whether it has been accessed recently or not, a modified bit M indicating whether it has been written into or not, and a frame number FN telling which main memory page frame contains the page. Every address generated by the CPU is decomposed into a page number part (i) and a line number part (x). The memory mapping unit (MMU) translates that address into a memory location as follows. It accesses memory location d+i, which contains the entry of page i in the page table PT[d]. If the page is present (P=1), it generates the memory location by substituting the frame number (f) for the page number (i). If it is not present (P=0), it instead generates a page fault interrupt that signals the operating system to invoke the page fault handler routine (PFH). The MMU also sets the use bit (U=1) and on write accesses the modified bit (M=1). The PFH selects a main memory page to replace, if modified copies it to the disk in its slot of the address space file AS[d], copies page i from the address space file to the empty frame, updates the page table, and signals the CPU to retry the previous instruction. As it searches for a page to replace, the PFH reads and resets usage bits, looking for unused pages. A copy of the most recent translations (from page to frame) is kept in the translation lookaside buffer (TLB), enabling the MMU to bypass the page table lookup most of the time.

algorithm could cost a million dollars of lost machine time over the life of a system, they all paid a great deal of attention to replacement algorithms.

Nonetheless, by 1966 these companies were reporting their systems were susceptible to a new, unexplained, catastrophic problem they called thrashing. Thrashing seemed to have nothing to do with the choice of replacement policy. It manifested as a sudden collapse of throughput as the multiprogramming level rose. A thrashing system spent most of its time resolving page faults and little running the CPU. Thrashing was far more damaging than a poor replacement algorithm. It scared the daylights out of the computer makers.

The more conservative IBM did not include virtual memory in its 360 operating system in 1964. Instead, it sponsored at its Watson laboratory one of the most comprehensive experimental systems projects of all time. Led by Bob Nelson, Les Belady, and David Sayre, the project team built the first virtual-machine operating system and used it to study the performance of virtual memory (The term "virtual memory" appears to have come from this project). In 1966 Belady published a landmark study in which he tested every replacement algorithm that anyone had ever proposed and a few more he invented. Many of his tests involved the use bits built into page tables (see Fig. 4.1) By periodically scanning and resetting the bits, the replacement algorithm distinguishes recently referenced pages from others. Belady concluded that policies favouring recently used pages performed better than other policies; LRU (least recently used) replacement was consistently the best performer among those tested[5].

4.3. Discovery and Propagation of Locality Idea (1966-1980)

In 1965, I entered my PhD studies at MIT in Project MAC, which was just undertaking the development of Multics. I was fascinated by the problems of dynamically allocating scarce CPU and memory resources among the many processes that would populate future time-sharing systems.

I set myself a goal to solve the thrashing problem and define an efficient way to manage memory with variable partitions. Solutions to these problems would be worth millions of dollars in recovered uptime of virtual memory operating systems. Little did I know that I would have to devise and validate a theory of program behaviour to accomplish this.

I learned about the controversies over the viability of virtual memory and was baffled by the contradictory conclusions among the experimental studies. All these studies examined individual programs assigned to a fixed memory partition managed by a replacement algorithm. They shed no light on the dynamic partitions used in multiprogrammed virtual memory systems. They offered no notion of a dynamic, intrinsic memory demand that would tell which pages of the program were essential and which were replaceable – something simple like, "this process needs p pages at time t." Such a notion was incompatible with the fixed-space policies everyone was studying. I began to speak of a process's intrinsic memory demand as its "working set". The idea was that paging would be acceptable if the system could guarantee that the working set was loaded. I combed the experimental studies looking for clues on how to measure a program's working set. All I could find were data on lifetime curves (mean time between page faults as a function of average memory space allocated to a program). These data suggested that the mean working set size would be significantly smaller than the full program size (Fig. 4.2).

In his 1966 study, Belady showed that clustering of references was fundamental to the performance of replacement algorithms. He used the term "locality" to characterise the clustering and showed that the FIFO algorithm systematically outperforms a random selection algorithm by capturing "locality sets" in memory. He found that LRU (least recently used) performed better than FIFO and introduced the use bit as a practical way to implement LRU. He invented the MIN algorithm, which minimised page faults by replacing the page with maximal forward time until next reference.

In an "Aha!" moment in the waning days of 1966, inspired by Belady's observations, I hit on the idea of defining a process's working set as the set of pages used during a fixed-length sampling window in the immediate past. A working set could be measured by periodically reading and resetting the use bits in a page table. The window had to be in the virtual time of the process – time as measured by the number of memory references made – so that the measurement would not be distorted by interruptions. This led to the now-familiar notation: the working set $W(t,T)$ is the set of pages referenced in the virtual time interval of length T preceding time t [14].

By spring 1967, I had an explanation for thrashing [15]. Thrashing was the collapse of system throughput triggered by making the multiprogramming level too high. It was counterintuitive because we were used to systems that would saturate under heavy load, not shut down (Fig. 4.3). When memory was filled with working sets, any further increment in the multiprogramming level would simultaneously push all loaded programs into a regime of working set insufficiency, where they paged excessively and could not use the CPU efficiently (Fig. 4.4). I proposed a feedback control mechanism that would limit the multiprogramming level by refusing to activate any program whose working set would not fit within the free space of main memory. When memory was full, the operating system would defer

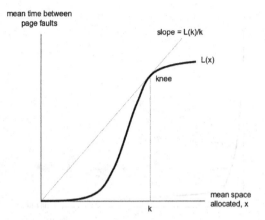

Fig. 4.2. A program's lifetime curve plots the mean time between page faults in a virtual memory system with a given replacement policy, as a function of the amount of space allocated to it by the system. It has an S-shape. The knee, defined as the point at which a line emanating from the origin is tangent to the curve, is the point of diminishing returns for increased memory allocation. The knee memory size is typically considerably smaller than the total program size, indicating that a replacement policy can often do quite well with a relatively small memory allocation. A further significance of the knee is that it maximises the ratio $L(x)/x$ for all points on the curve. The knee is therefore the most desirable target for space allocation: it maximises the mean time between faults per unit of space.

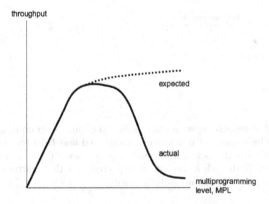

Fig. 4.3. A computer system's throughput (jobs completed per second) increases with multiprogramming level up to a point. Then it decreases rapidly to throughput so low that the system appears to have shut down. Because everyone was used to systems that gradually approach saturation with increasing load, the throughput collapse was unexpected. The thrashing state was "sticky" – we had to reduce the MPL somewhat below the trigger point to get the system to reset. No one knew how to predict the optimal MPL or to find it without falling into thrashing.

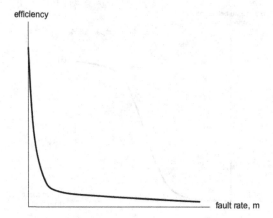

Fig. 4.4. The first rigorous explanation of thrashing argued from efficiency. The efficiency of a program is the ratio of its CPU execution time to its real time. Real time is longer because of page-fault delays. Denote a program's execution time by E, the page fault rate by m, and the delay for one page fault by D; then the efficiency is $E/(E+mED) = 1/(1+mD)$. For typical values of D – 10,000 memory cycle times or longer – the efficiency drops very rapidly for a small increase of m above 0. In a memory filled with working sets (high efficiency), loading one more program can squeeze all the others, pushing everyone into working set insufficiency, collapsing efficiency.

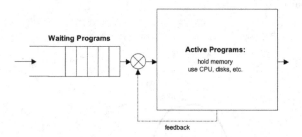

Fig. 4.5. A feedback control system can stabilise the multiprogramming level and prevent thrashing. The amount of free space is monitored and fed back to the scheduler. The scheduler activates the next waiting program whenever the free space is sufficient for its working set. With such a control, we expected that the multiprogramming level would rise to the optimal level and stabilise there.

programs requesting activation into a holding queue. Thrashing would be impossible with a working set policy (Fig. 4.5).

The working set idea was based on an implicit assumption that the pages seen in the backward window were highly likely to be used again in the immediate future. Was this assumption justified? In discussions with Jack Dennis (MIT) and Les Belady (IBM), I started using the term "locality"

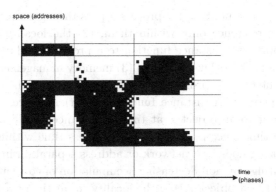

Fig. 4.6. Locality sequence behaviour observed by sampling use bits during program execution. Programs exhibit phases and localities naturally, even when overlays are not pre-planned.

for the observed tendency of programs to cluster references to small subsets of their pages for extended intervals[19]. We could represent a program's memory demand as a sequence of locality sets and their holding times:

$$(L1, T2), (L2, T2), (L3, T3), \ldots, (Li, Ti), \ldots$$

This seemed natural because we knew that programmers planned overlays using diagrams that showed subsets and time phases. But what was strikingly interesting was that programs showed the locality behaviour even when it was not explicitly pre-planned. When measuring actual page use, we repeatedly observed many long phases with relatively small locality sets (Fig. 4.6). Each program had its own distinctive pattern, like a voiceprint.

We saw two reasons that this would happen: (1) temporal clustering due to looping and executing within modules with private data, and (2) spatial clustering due to related values being grouped into arrays, sequences, modules, and other data structures. Both these reasons seemed related to the human practice of "divide and conquer" – breaking a large problem into parts and working separately on each. The locality bit maps captured someone's problem-solving method in action. These underlying phenomena gave us confidence to claim that programs have natural sequences of locality sets. The working set sequence is a measurable approximation of a program's intrinsic locality sequence.

As we developed and refined our understanding of locality during the 1970s, I continued to work with many others to refine the locality idea and turn it into a behavioural theory of computational processes interacting with storage systems. By 1980 we articulated the principle as a package of three ideas: (1) computational processes pass through a sequence of locality sets and reference only within them, (2) the locality sets can be inferred by applying a distance function to a program's address trace observed during a backward window, and (3) memory management is optimal when it guarantees each program that its locality sets will be present in high-speed memory[17]. A distance function $D(x, t)$ measures the distance from a processor to an object x at time t. Distances could be temporal, measuring the time since prior reference or access time within a network; spatial, measuring hops in a network or address separation in a sequence; or cost, measuring any non-decreasing accumulation of cost since prior reference. We said that object x is in the locality set at time t if the distance is less than a threshold: $D(x, t) \leq T$. The storage system would maximise throughput by caching locality sets close to the processor.

By 1975, the queuing network model had become a useful tool for understanding the performance of computing systems, and for predicting throughput, response time, and system capacity. In this model, each computing device of the real system is represented as a server with a queue; the server processes a job for a random service time and then sends it to another server according to a probability distribution for the inter-server transition. The parameters of the model are the mean service times for each server, the mean number of times a job visits a server, and the total number of jobs circulating in the system. We began to use these models to study how to tell when a computing system had achieved its maximum throughput and was on the verge of thrashing. The results were eye-opening.

In the simplest queuing model of a virtual memory system, there is a server representing the CPU and a server representing the paging disk. A job cycles between the CPU and the disk in the pattern

$(CPU, Disk)^*CPU$

meaning a series of CPU-Disk cycles followed by a CPU interval before completing. The number of CPU-Disk cycles is the number of page faults generated by the system's replacement policy for the mean memory space allocated to jobs. Queuing network theory told us that every server poses a potential bottleneck that imposes an upper limit on the system throughput; the actual bottleneck is the server with the smallest limit. We discovered that the well-known thrashing curve (Fig. 4.3) is actually the system doing the best it can as the paging-disk bottleneck worsens with increasing load (Fig. 4.7).

Once we saw that thrashing is a bottleneck problem, we studied whether we could use bottleneck parameters as criteria for load controls that

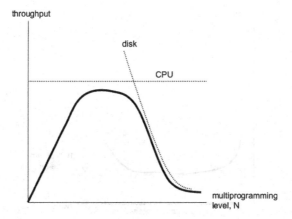

Fig. 4.7. System throughput is constrained by both CPU and disk capacity. The CPU imposes a throughput limit of $1/R$, where R is the average running time of programs. The disk imposes a throughput limit of $1/SF$, where S is the mean time to do a page swap and F is the total number of page faults in a job. Thrashing is caused by precipitous drop of disk capacity as increased load squeezes space and forces more paging. The crossing point occurs when $R = SF$; since $F = R/L$ (lifetime, L), the crossing is at $L = S$, i.e., when the mean time between faults equals the disk service time of a fault. Thus a control criterion is to allow N to increase until L decreases to S. Unfortunately, this was not very precise; we found experimentally that many systems were already in thrashing when $L = S$. Moreover, the memory size at which $L = S$ may bear no relation to the highly desirable lifetime knee (Fig. 4.2).

prevented thrashing. One such criterion was called "$L = S$" because it involved monitoring the mean lifetime L between page faults and adjusting load to keep that value near the paging disk service time S (Fig. 4.7). This criterion was not very reliable: in some experiments, the system would already be thrashing when $L = S$. We found that a "knee criterion" – in which the system adjusted load to keep the observed lifetime near the knee lifetime (Fig. 4.2) – was consistently more reliable, even though knee lifetime was not close to S. Unfortunately, it is not possible to know the knee lifetime without running the program to completion.

Our theory told us that system throughput would be maximum when space-time for each job is minimum, confirming our claim that a knee criterion would optimise throughput. How well can a working-set policy approach this ideal? In a line of experimental studies we found that the interval of window values that put the space-time within 10% of its minimum was quite wide (Fig. 4.8)[27]. Then we found that many workloads, consisting of a variety of programs, often had global T values that fell in all the 10%

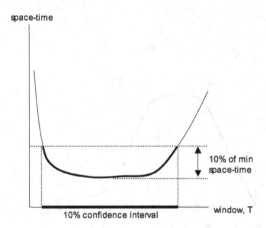

Fig. 4.8. System throughput is maximised when the memory space-time consumed by a job is minimum. The memory allocation that does this is near the knee (Fig. 4.2). Our experimental studies of working-set windows near the knee of its lifetime curve yielded two useful results. One is that a program's space-time is likely to be flat (near minimum) for a broad range of window sizes. the picture shows how we define a 10% confidence interval of window sizes.

confidence intervals. This meant that a single, fixed, properly chosen value of T would cause the working set policy to maintain system throughput to within 10% of its optimum. The average deviation was closer to 5%. The conclusion was that systems with a properly adjusted, single global T value would achieve a working-set throughput within 5-10% of optimal.

The final question was: is there another policy that would deliver a lower space-time per job and therefore a higher optimum throughput? Obviously, the VMIN (variable space minimum[37]) would do the job; but it requires lookahead. We discovered that the working set policy has exactly the same page-fault sequence as VMIN. Therefore the difference of space-time between WS and VMIN is completely explained by working-set "overshooting" in its estimates of locality at the transitions between program phases. Indeed, VMIN unloads pages toward the ends of their phases after it sees they will not be referenced in the next phase. Working set cannot tell this until time T after the last reference. Experiments by Alan Smith to clip off these overshoots showed only a minor gain[42]. We concluded that it would be unlikely that anyone would find a non-lookahead policy that was noticeably better than working set.

Thus, by 1976, our theory was validated. It demonstrated our original postulate: that working set memory management would prevent thrashing and would allow system throughput to be close to its optimum.

The problem of thrashing, which originally motivated the working set theory, has occurred in other contexts as well as storage management. It can happen in any system where contention for a shared resource can overwhelm the processes' abilities to move forward. It was observed in the first packet-radio communication system, ALOHA, in the late 1960s. In this system, the various contenders could overwhelm the shared spectrum by retransmitting packets when they discovered their transmissions being inadvertently jammed by other transmitters[1]. A similar problem occurred in the Ethernet, where it was solved by the "back-off" protocol that makes a transmitter wait a random time before retrying a transmission[35]. A similar problem occurred in database systems with the two-phase commit protocol[45]. Under this protocol, transactions try to collect locks on the records they will update; but if they find any record already locked, they release all their locks and try again. When too many transactions try to collect locks at the same time, they spend most of their time gathering and releasing locks.

Although it is not critical to the theory and conclusions above, it is worth noting that the working-set analysis applies even when processes share pages. Among its design objectives, Multics supported multiprocess (multithreaded) computations. The notions of locality and working sets had to apply in this environment. The obvious approach was to define a computation's working set as the union of its constituent process working sets. This approach did not work well in the standard paging system architecture (Fig. 4.1) because the use bits that had to be OR'd together were in different page tables and a high overhead would be needed to locate them. Fortunately, the idea of capability-based addressing, a locality-enhancing architecture articulated by colleagues Dennis and Van Horn in 1966[22], offered a solution (Fig. 4.9). Working sets could be measured from the use bits of the set of object descriptors.

The two-level mapping inherent in capability addressing is a principle in its own right. It solved a host of sharing problems in virtual memories of multiprocess operating systems[23]. It stimulated a line of extraordinarily fault tolerant commercial systems known as "capability machines" in the 1970s[47,49]. The architecture was adopted into the run time environments of object oriented programming systems. The principle was applied to solving the problem of sharing objects in the Internet[29]. Thus the situations in which working sets and localities of multithreaded and distributed computations apply are ubiquitous today.

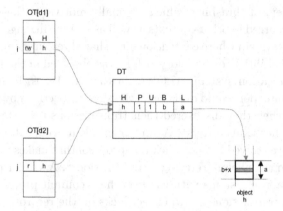

Fig. 4.9. Two-level mapping enables sharing of objects without prior arrangements among the users. It begins with assigning a unique (over all time) identifying handle h to an object; objects can be of any size. Object h has a single descriptor specifying its status in memory: present ($P = 0$ or 1), used ($U = 0$ or 1), base address (B, defined only when $P = 1$), and length (L). The descriptors of all known objects are stored in a descriptor table DT, a hash table with the handle as a lookup key. When present, the object is assigned a block of contiguous addresses in main memory. Each computational process operates in its own memory domain (such as $d1$ or $d2$), which is specified by an object table (OT), an adaptation of the page table (Fig. 4.1). The object table, indexed by an object number (such as i or j), retrieves an object's access code (such as rw) and handle. The memory mapping unit takes an address (i,x), meaning line x of object i, and retrieves the handle from the object table; then it looks up the descriptor for the handle in the descriptor table; finally it forms the actual memory address $b + x$ provided that x does not exceed the object's size a. Any number of processes can share h, simply by adding entries pointing to h as convenient in their object tables. Those processes can use any local name (i or j) they desire. If the system needs to relocate the object in memory, it can do so by updating the descriptor (in the descriptor table). All processes will get the correct mapping information immediately. Working sets can be measured from the use bits (U) in the descriptor table.

Appendix 1 summarises milestones in the development of the locality idea.

4.4. Adoption of Locality Principle (1967-present)

The locality principle was adopted as an idea almost immediately by operating systems, database, and hardware architects. But it did not remain a pure idea for long. It was adopted into practice, in ever widening circles:

- In virtual memory to organise caches for address translation and to design the replacement algorithms.
- In data caches for CPUs, originally as mainframes and now as microchips.

- In buffers between main memory and secondary memory devices.
- In buffers between computers and networks.
- In video boards to accelerate graphics displays.
- In modules that implement the information-hiding principle.
- In accounting and event logs in that monitor activities within a system.
- In alias lists that associate longer names or addresses with short nicknames.
- In the "most recently used" object lists of applications.
- In web browsers to hold recent web pages.
- In search engines to find the most relevant responses to queries.
- In classification systems that cluster related data elements into similarity classes.
- In spam filters, which infer which categories of email are in the user's locality space and which are not.
- In "spread spectrum" video streaming that bypasses network congestion and reduces the apparent distance to the video server.
- In "edge servers" to hold recent web pages accessed by anyone in an organisation or geographical region.
- In the field of computer forensics to infer criminal motives and intent by correlating event records in many caches.
- In the field of network science by defining hierarchies of self-similar locality structures within complex power-law networks.

Appendix 2 summarises milestones in the adoption of locality in systems. The locality principle is today considered as a fundamental principle for systems design.

4.5. Modern Model of Locality: Context Awareness

As the uses of locality expanded into more areas, our understanding of locality has evolved beyond the original idea of clustering of reference. Today's understanding embraces four key ideas that enable awareness of, and meaningful response to, the context in which software is situated (Fig. 4.10):

- **An observer**;
- **Neighbourhoods:** One or more sets of objects that are most relevant to the observer at any given time;
- **Inference:** A method of identifying the most relevant objects by monitoring the observer's actions and interactions and other infor-

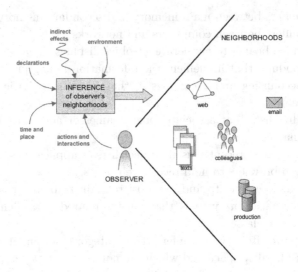

Fig. 4.10. The modern view of locality is a means of inferring the context of an observer using software, so that the software can dynamically adapt its actions to produce optimal behaviour for the observer.

mation about the observer contained in the environment; and
- **Optimal actions:** An expectation that the observer will complete work in the shortest time if neighbourhood objects are ready accessible in nearby caches.

The observer is the agent who is trying to accomplish tasks with the help of software, and who places expectations on its function and performance. In most cases, the observer is the user who interacts with software. In some cases, especially when a program is designed to compute a precise, mathematical model, the observer can be built into the software itself.

A neighbourhood is a group of objects standing in a particular relation to an observer, and valued in some way by the observer. The historical example is the locality set associated with each phase of a program's execution; the observer is the process evoked by the program. Modern examples of neighbourhoods include email correspondents, non-spam email, colleagues, teammates, objects used in a project, favourite objects, user's web, items of production, texts, and directories.

Some neighbourhoods can be known by explicit declarations; for example a user's file directory, address book, or web pages. But most neighbourhoods can only be inferred by monitoring the event sequences of an observer's actions and interactions. The event sequences can be measured either within the software with which the observer is interacting, or outside that software, in the run-time system.

Inference can be any reasonable method that estimates the content of neighbourhoods, or at least the most relevant parts. The original working estimated locality sets by observing page reference events in a backward looking window. Modern inference is by a variety of methods such as Google's counting of incoming hyperlinks to a web page, connectionist networks that learn patterns after being presented with many examples, or Bayesian spam filters.

The optimal actions are taken by the system or the software on behalf of the observer. As with the data collection to support neighbourhood inference, these actions can come either from inside the software with which the observer is interacting, or from outside that software, in the run-time system. The matrix below shows four quadrants corresponding to the four combinations of data collection and locus of action just mentioned.

		Origin of Data for Inference	
		Inside	Outside
Locus of Adaptive Action	Inside	Amazon.com, Bayesian spam filter	Semantic web Google
	outside	Linkers and loaders	Working sets, Ethernet load control

Examples of software are named in each quadrant and are summarised below.

- **Amazon.com; Bayesian spam filters.** This system collects data about user purchasing histories and recommends other purchases that resemble the user's previous purchases, or purchases by other, similar users. Bayesian spam filters gather data about which emails the user considers relevant and then block irrelevant emails. (Data collection inside, optimal actions inside.)
- **Semantic web; Google.** Semantic web is a set of declarations of structural relationships that constitute context of objects and their connections. Programs read and act on it. Google gathers data from the Web and uses it to rank pages that appear to be most relevant to a keyword query posed by user. (Data collection outside, optimal actions inside.)

- **Linkers and Loaders.** These workhorse systems gather library modules mentioned by a source program and link them together into a self-contained executable module. The libraries are neighbourhoods of the source program. (Data collection inside, optimal action outside.)
- **Working sets, Ethernet load controls.** Virtual memory systems measure working sets and guarantee programs enough space to contain them, thereby preventing thrashing. Ethernet prevents the contention resolving protocol from getting overloaded by making competing transactions wait longer for retries if load is heavy. (Data collection outside, optimal action outside.)

In summary, the modern principle of locality is that observers operate in one or more neighbourhoods that can be inferred from dynamic action sequences and static structural declarations. Systems can optimise the observer's productivity by adapting to the observer's neighbourhoods, which they can estimate by distance metrics or other inferences.

4.6. Future Uses of Locality Principle

Locality principles are certain to remain at the forefront of systems design, analysis, and performance. This is because locality flows from human cognitive and coordinative behaviour. The mind focuses on a small part of the sensory field and can work most quickly on the objects of its attention. People organise their social and intellectual systems into neighbourhoods of related objects, and they gather the most useful objects of each neighbourhood close around them to minimise the time and work of using them. These behaviours are transferred into the computational systems we design and into the expectations users have about how their systems should interact with them.

Here are seven modern areas offering challenging research problems that locality may be instrumental in solving.

Architecture. Computer architects have heavily exploited the locality principle to boost the performance of chips and systems. Putting cache memory near the CPU, either on board the same chip or on a neighbouring chip, has enabled modern CPUs to pass the 1 GHz speed mark. Locality within threaded instruction sequences is being exploited by a new generation of multi-core processor chips. The "system on a chip" concept places neighbouring functions on the same chip to significantly decrease delays of communicating between components. Locality is used to compress animated sequences of pictures by detecting the common neighbourhood behind a sequence and transmitting it once and then transmitting the differences. Architects will continue to examine locality carefully to find new ways to speed up chips, communications, and systems.

Caching. The locality principle is useful wherever there is an advantage in reducing the apparent distance from a process to the objects it can access. Objects in the process's neighbourhood are kept in a local cache with fast access time. The performance acceleration of a cache generally justifies the modest investment in the cache storage. Novel forms of caching have sprung up in the Internet. One prominent example is edge servers that store copies of web objects near their users. Another example is the clustered databases built by search engines (like Google) to instantly retrieve relevant objects from the same neighbourhoods as the asker. Similar capabilities are available in MacOS 10.4 (Tiger) and will be in Windows 2006 to speed up finding relevant objects.

Bayesian Inference. A growing number of inference systems exploit Bayes's principle of conditional probability to compute the most likely internal (hidden) states of a system given observable data about the system. Spam filters, for example, use it to infer the email user's mental rules for classifying certain objects as spam. Connectionist networks use it for learning: the internal states that abstract from desired input-output pairs shown to the network; the network gradually acquires a capability for new action. Bayesian inference is an exploitation of locality because it infers a neighbourhood given observations of what a user or process is doing.

Forensics. The burgeoning field of computer forensics owes much of its success to the ubiquity of caches. They are literally everywhere in an operating systems and applications. By recovering evidence from these caches, forensics experts can reconstruct (infer) an amazing amount of a criminal's motives and intent[24]. Criminals who erase data files are still not safe because experts use advanced signal-processing methods to recover the faint magnetic traces of the most recent files from the disk[11]. Learning to draw valid inferences from data in a computer's caches, and from correlated data in caches in other computers with which the subject has communicated, is a challenging research problem.

Web Based Business Processes. The principle of locality has pervaded the design of web based business systems, which allow buyers and sellers to engage in transactions using web interfaces to sophisticated database systems. Amazon.com illustrates how a system can infer "book interest neighbourhoods" of customers and (successfully) recommend additional sales. Many businesses employ customer relationship management (CRM) systems that infer "customer interest neighbourhoods" and allow the company to provide better, more personalised service. Database, network, server, memory, and other caches optimise the performance of these systems[34].

Context Aware Software. A growing number of software designers are coming to believe that most software failures can be traced to the inability of software to be aware of and act on the context in which it operates. The modern locality principle is beginning to enable software designers to reconstruct context and thereby to be consistently more reliable, dependable, usable, safe, and secure.

Network Science. Inspired by A. L. Barabasi, many scientists have begun applying statistical mechanics to large random networks, typically finding that the distribution of node connections is power law with degree -2 to -3 in most cases[4,18]. These networks have been found to be self-similar, meaning that if all neighbourhoods (nodes within a maximum distance of each other) are collapsed to single nodes, the resulting network has the same power distribution as the original[43]. The idea that localities are natural in complex systems is not new; in 1976 Madison and Batson reported that program localities have self-similar sub-localities[32], and in 1977 P. J. Courtois applied it to cluster similar states of complex systems to simplify their performance analyses[13]. The locality principle may offer new understandings of the structure of complex networks.

Researchers looking for challenging problems can find many in these areas and can exploit the principle of locality to solve them.

References

1. Abramson, N., "The ALOHA System–Another Alternative for Computer Communication" 1970, 281-285. (Proc. AFIPS Fall Joint Computer Conference 37, 1970)
2. Aho, A. V., P. J. Denning, and J. D. Ullman., "Principles of optimal page replacement" *J ACM* **18**, (Jan 1971), 80-93.
3. Badel, M., E. Gelenbe, J. Lenfant, and D. Potier., "Adaptive optimization of a time sharing system's performance" *Proc. IEEE* **63**, (1975) 958-965.
4. Barabasi, A. L., *Linked: The New Science of Networks*, (Perseus Books), 2002.
5. Belady, L. A., "A study of replacement algorithms for virtual storage computers" *IBM Systems J.* **5**(2), (1966), 78-101.
6. Blake, R., "Optimal control of thrashing" 1984, 1-10. (ACM SIGMETRICS Proc. Conf. on Measurement and Modeling of Computer Systems, 1984)
7. Brawn, B., and F. G. Gustavson., "Program behavior in a paging

environment" Thompson 1968, 1019-1032. (Proc. AFIPS Fall Joint Computer Conference 33, 1968)
8. Buzen, J. P., "Optimising the degree of multiprogramming in demand paging systems", 1971, 139-140 (Proc. IEEE COMPCON, Sep 1971).
9. Buzen, J. P., "Fundamental laws of computer systems performance" *Acta Informatica* **7**(2), (1976), 167-182.
10. Carr, J., and J. Hennessy., "WSCLOCK – a simple and effective algorithm for virtual memory management" 1981, 87-95. (ACM SIGOPS Proc. 8th Symp. on Operating Systems Principles, 1981)
11. Carrier, Brian., *File System Forensic Analysis*, (Addison Wesley), 2005.
12. Corbato, F. J., "A paging experiment with the Multics system" In Honor of P. M. Morse, K. U. Ingard, Ed. MIT Press, (1969), 217-228.
13. Courtois, P. J., *Decomposability*, (Academic Press), 1977.
14. Denning, P. J., "The working set model for program behavior" *ACM Communications* **11**(5), (May 1968), 323-333.
15. Denning, P. J., "Thrashing: Its causes and prevent" Thompson 1968, 915-922. (Proc. AFIPS Fall Joint Computer Conference 33, 1968)
16. Denning, P. J., "Virtual memory" *ACM Computing Surveys* **2**(3), (Sept 1970), 153-189.
17. Denning, P. J., "Working sets past and present" *IEEE Transactions on Software Engineering* **6**(1), (January 1980), 64-84.
18. Denning, P. J., "Network Laws" *ACM Communications* **47**, (Nov 2004), 15-20.
19. Denning, P. J., and S. C. Schwartz., "Properties of the working set model" *ACM Communications* **15**, (March 1972), 191-198.
20. Denning, P. J., K. C. Kahn, J. Laroudier, D. Potier, and R. Suri., "Optimal multiprogramming" *Acta Informatica* **7**(2), (1976), 197-216.
21. Denning, P. J., and D. R. Slutz., "Generalised working sets for segment reference strings" *ACM Communications* **21**(9), (September 1978), 750-759.
22. Dennis, J. B., and E. C. van Horn., "Programming semantics for multiprogrammed computations" *ACM Communications* **9**(3), (March 1966), 143-155.
23. Fabry, R. S., "Capability-Based Addressing" *ACM Communications* **17**(7), (July 1974), 403-412.
24. Farmer, D., and W. Venema. *Forensic Discovery*, (Addison Wesley), 2004.
25. Ferrari, D., "Improving locality by critical working sets" *ACM Communications* **17**, (Nov 1974), 614-620.
26. Gelenbe, E., J. Lenfant, and D. Potier., "Analyse d'un algorithme de gestion de memoire centrale et d'un disque de pagination" *Acta Informatica* **3**, (1974), 321-345.
27. Graham, G. S., and P. Denning., "Multiprogramming and program behavior" 1974, 1-8. (Proc ACM SIGMETRICS Conference on Measurement and Evaluation, 1974))
28. Hatfield, D., and J. Gerald., "Program restructuring for virtual memory" *IBM Syst. J.* **10**, (1971), 168-192.
29. Kahn, R., and R. Wilensky., "A framework for distributed digital object

services" Corporation for National Research Initiatives, (1995) http://www.cnri.reston.va.us/k-w.html.
30. Kilburn, T., D. B. G. Edwards, M. J. Lanigan, F. H. Sumner., "One-level storage system" *IRE Transactions* **EC-11**, (April 1962), 223-235.
31. Leroudier, J., and D. Potier., "Principles of optimality for multiprogramming" 1976, 211-218. (Proc. Int'l Symp. Computer Performance Modeling, Measurement, and Evaluation, ACM SIGMETRICS and IFIP WG 7.3, March 1976)
32. Madison, A. W., and A. Batson., "Characteristics of program localities" *ACM Communications* **19**(5), (May 1976), 285-294.
33. Mattson, R. L., J. Gecsei, D. R. Slutz, I. L. Traiger., "Evaluation techniques for storage hierarchies" *IBM Systems J.* **9**(2), (1970), 78-117.
34. Menasce, D., and V. Almeida., *Scaling for E-Business: Technologies, Models, Performance, and Capacity Planning*, (Prentice-Hall), 2000.
35. Metcalfe, R. M., and D. Boggs., "Ethernet: Distributed packet switching for local networks" *ACM Communications* **19**(7), (July 1976), 395-404.
36. Morris, J. B., "Demand paging through the use of working sets on the MANIAC II" *ACM Communications* **15**, (Oct 1972), 867-872.
37. Prieve, B., and R. S. Fabry., "VMIN – An optimal variable space page replacement algorithm" *ACM Communications* **19**, (May 1976), 295-297.
38. Roberts, L. G., "ALHOA packet system with and without slots and capture" *ACM SIGCOMM Computer Communication Review* **5**(2), (April 1975), 28-42.
39. Rodriguez-Rosell, J., and J. P. Dupuy., "The design, implementation, and evaluation of a working set dispatcher" *ACM Communications* **16**, (Apr 1973), 247-253.
40. Sayre, D., "Is automatic folding of programs efficient enough to displace manual?" *ACM Communications* **13**, (Dec 1969), 656-660.
41. Shore, J. E., "The lazy repairman and other models: performance collapse due to overhead in simple, single-server queueing systems" 1982, 217-224. (ACM SIGMETRICS Proc. Intl. Symp. on Performance Measurement, Modeling, and Evaluation, May 1982)
42. Smith, A. J., "A modified working set paging algorithm" *IEEE Trans. Computing* **25**, (Sep 1976), 907-914.
43. Song, C., S. Havlin, and H. Makse., "Self-Similarity of Complex Networks" *Nature* **433**, (Jan 2005), 392-395.
44. Spirn, J., *Program Behavior: Models and Measurements*, (Elsevier Computer Science), 1977.
45. Thomasian, A., "Two-phase locking performance and its thrashing behavior" *ACM Trans. Database Systems* **18**(4), (December 1993), 579-625.
46. Wilkes, M. V., "Slave memories and dynamic storage allocation" *IEEE Transactions Computers* **14**, (April 1965), 270-271.
47. Wilkes, M. V., *Time Sharing Computer Systems*, (Elsevier), 1972.
48. Wilkes, M. V., "The dynamics of paging" *Computer J* **16**, (Feb 1973), 4-9.
49. Wilkes, M. V., and R. Needham., *The Cambridge CAP Computer and Its Operating System*, (Elsevier North-Holland), 1979.

Appendix 1 – Milestones in Development of Locality Idea

1959	Atlas operating system includes first virtual memory; a "learning algorithm" replaces pages referenced farthest in the future [30].
1961	IBM Stretch supercomputer uses spatial locality to prefetch instructions and follow possible branches.
1965	Wilkes introduces slave memory, later known as CPU cache, to hold most recently used pages and significantly accelerate effective CPU speed [46].
1966	Belady at IBM Research publishes comprehensive study of page replacement algorithms, showing that those with use bits outperform others and proposing an initial definition of locality to explain why [5]. Corbato reconfirms for Multics [12].
1966	Denning proposes working set idea: the pages that must be retained in main memory are those referenced during a window of length T preceding the current time. In 1967 he postulates that working set memory management will prevent thrashing [14,15,17].
1968	Denning shows analytically why thrashing precipitates suddenly with any increase above a critical threshold of number of programs in memory [15]. Belady and Denning use term locality for the program behaviour property working sets measure.
1969	Sayre, Brawn, and Gustavson at IBM demonstrate that programs with good locality are easy to design and cause virtual memory systems to perform better than a manually design paging schedule [7, 40].
1970	Denning gathers all extant results for virtual memory into Computing Surveys paper "virtual memory" that was widely used in operating systems courses. This was first coherent scientific framework for designing and analysing dynamic memories [16].
1970-71	Mattson, Gecsei, Slutz, and Traiger of IBM publish "stack algorithms", modelling a large class of popular replacement policies including LRU and MIN and offering surprisingly simple algorithms for calculating their paging functions in virtual memory [33]. Aho, Denning, and Ullman prove a principle of optimality for page replacement [2].
1971	Hatfield and Gerald demonstrate compiler code generation methods for preserving locality in executable files [28]. Ferrari shows even greater gains when working sets measure locality [25].
1972	Spirn and Denning conclude that locality sequence (phase-transition) behaviour is the most accurate description of locality [44].
1970-74	Abramson, Metcalfe, and Roberts report thrashing in Aloha and Ethernet communication systems; load control protocols prevent it [1,35,38].
1976	Buzen, Courtois, Denning, Gelenbe, and others integrate memory management into queuing network models, demonstrating that thrashing is caused by the paging disk transitioning into the bottleneck with increasing load [3,8,9,13,20,26,31]. System throughput is maximum when the average working set space-time is minimum [9,27].
1976	Madison and Batson demonstrate that locality is present in symbolic execution strings of programs, concluding that locality is part of human cognitive processes transmitted to programs [32]. They show that locality sequences have self-similar substructures.

1976	Prieve and Fabry demonstrate VMIN, the optimal variable-space replacement policy [37]; it has identical page faults as working set but lower space-time accumulation at phase transitions [17].
1978	Denning and Slutz define generalised working sets; objects are local when their memory retention cost is less than their recovery costs. The GWS models the stack algorithms, space-time variations of working sets, and all variable-space optimal replacement algorithms [21].
1980	Denning gathers the results of over 200 virtual-memory researchers and concludes that working set memory management with a single system-wide window size is as close to optimal as can practically be realised [17].
1981	Carr and Hennessy offer effective software implementation of working set by applying sampling windows in CLOCK algorithm [10].
1982-84	Shore reports thrashing in large class of queuing systems [41]. Blake offers optimal controls of thrashing [6].
1993	Thomasian reports thrashing in two-phase locking systems [45].

Appendix 2 – Milestones in Adoption of Locality

1961	IBM Stretch computer uses spatial locality for instruction lookahead.
1964	Major computer manufacturers (Burroughs, General Electric, RCA, Univac but not IBM) introduce virtual memory with their "third generation computing systems". Thrashing is a significant performance problem.
1965-1969	Nelson, Sayre, and Belady, at IBM Research built first virtual machine operating system; they experiment with virtual machines, contribute significant insights into performance of virtual memory, mitigate thrashing through load control, and lay groundwork for later IBM virtual machine architectures.
1968	IBM introduces cache memory in 360 series. Multics adopts "clock", an RLU variant, to protect recently used pages.
1969-1972	Operating systems researchers demonstrate experimentally that the working set policy works as advertised. They show how to group code segments on pages to maximise spatial locality and thus temporal locality during execution [36, 39, 48].
1972	IBM introduces virtual machines and virtual memory into 370 series. Bayer formally introduces B-tree for organising large files on disks to minimise access time by improving spatial locality. Parnas introduces information hiding, a way of localising access to variables within modules.
1978	First BSD Unix includes virtual memory with load controls inspired by working set principle; propagates into Sun OS (1984), Mach (1985), and Mac OS X (1999).
1974-79	IBM System R, an experimental relational database system, uses LRU managed record caches and B-trees.
1981	IBM introduces disk controllers containing caches so that database systems can get records without a disk access; controllers use LRU but do not cache records involved in sequential file accesses.
early 1980s	Chip makers start providing data caches in addition to instruction caches, to speed up access to data and reduce contention at memory interface.
late 1980s	Application developers add "most recent files" list to desktop applications, allowing users to more quickly resume interrupted tasks.
1987-1990	Microsoft and IBM develop OS/2 operating systems for PCs, with full multitasking and working set managed virtual memory. Microsoft splits from IBM, transforms OS/2 into Windows NT.
Early 1990s	Computer forensics starts to emerge as a field; it uses locality and signal processing to recover the most recently deleted files; and it uses multiple system and network caches to reconstruction actions of users.
1990-1998	Beginning with Archie, then Gopher, Lykos, Altavista, and finally Google, search engines compile caches that enable finding relevant documents from anywhere in the Internet very quickly.
1993	Mosaic (later Netscape) browser uses a cache to store recently accessed web pages for quick retrieval by the browser.
1995	Kahn and Wilensky show a method of digital object identifiers based on the locality-enhancing two-level address mapping principle.
1998	Akamai and other companies provide local web caches ("edge servers") to speed up Internet access and reduce traffic at sources.

CHAPTER 5

A Simulation-Based Performance Analysis of Epoch Task Scheduling in Distributed Processors

Helen Karatza

Department of Informatics, Aristotle University of Thessaloniki
54124 Thessaloniki, Greece
E-mail: karatza@csd.auth.gr

Efficient scheduling of parallel jobs on distributed processors is essential for good performance. In this article parallel jobs that consist of independent tasks are considered and a special type of task scheduling referred to as *epoch task scheduling* is studied. With this policy processor queues are rearranged according to the Shortest Task First (STF) method only at the end of predefined intervals. The time interval between successive queue rearrangements is called an *epoch*. The objective is to examine if epoch scheduling can perform well as compared to the STF method, i.e. reduce significantly the number of queue rearrangements and achieve fairer service than that of STF. A simulation model is used to address performance issues associated with epoch scheduling. Simulated results indicate that there are cases where epoch scheduling can succeed in these pursuits.

5.1. Introduction

It is still not clear how to efficiently schedule parallel jobs in distributed systems. To schedule them, it is necessary to divide programs into tasks, assign the tasks to distributed processors and then schedule them on processors. Good scheduling policies can maximise both system and application performance, and avoid unnecessary delays. Many papers study distributed system scheduling algorithms. The reference section lists a few of them[1-18].

Most research into distributed system scheduling policies focuses on improving performance metrics where scheduling overhead is assumed to be negligible. However, the overhead can seriously degrade performance. Also, fairness among individual jobs is often not considered.

FCFS (First Come First Served) is the simplest scheduling algorithm. It is fair to individual jobs but often produces sub-optimal performance and it incurs no overhead. Many scheduling algorithms have been proposed to achieve higher performance by considering information regarding individual requests.

STF (Shortest Task First or Shortest Time First) is a non-preemptive algorithm that minimizes average delay in the case of single task jobs. One disadvantage of STF is unfairness: some tasks may starve or suffer inordinate queuing delays particularly at high system loads. Since short tasks are given priority over long tasks, it is possible to starve a task if its service time is very large in comparison to the mean service time.

When parallel jobs consisting of independent tasks are assigned to distributed processors, the performance of STF method varies widely depending on the workload. This is because STF may assign different priorities to tasks within the same job. This can cause large synchronisation delays among sibling tasks[4,7].

To implement STF, *a priori* knowledge about task service times is needed. This requirement causes implementation problems since *a priori* knowledge is often not available in practice. However, policies like this are often studied to determine how much benefit is possible if service times are known in advance. STF is often used as a yardstick to compare performance of other scheduling policies.

Another disadvantage of STF is that it requires a considerable amount of overhead because processor queues are rearranged each time new tasks are added.

To address the problems of unfairness and queue rearrangement overhead, this article studies a version of STF called Epoch STF scheduling. Processor queues are rearranged only at the end of epochs, and then the scheduler recalculates the priorities of all tasks in the system queues using the STF criterion.

This type of scheduling is different from epoch scheduling that is studied in McCann and Zahorjan[14] where co-scheduling is considered and all nodes are reallocated to jobs at each reallocation point.

Other papers that study epoch scheduling include Karatza[8-11]. In Karatza[8] a closed queuing network model is considered with various degrees of multiprogramming, while in this paper an open queuing network model is studied with various job arrival rates. In Karatza[9] tasks of a job perform clusters at different processors, where the tasks in each cluster have to be executed in sequence one after the other. In Karatza[9] cluster scheduling is examined. In this chapter, all tasks of a job are independent without any precedence constraints and scheduling occurs at the task level. In Karatza[10] jobs are not parallel and therefore scheduling occurs at the job level. Epoch scheduling of parallel jobs consisting of independent tasks is also examined in Karatza[11] where the probabilistic task routing is employed and tasks with exponential service demands are considered. This research is an extension of Karatza[11] in that it also employs the "join the shortest queue" routing criterion and furthermore it examines tasks with high service demand variability.

The goal of this research is to investigate whether epoch STF scheduling can combine good performance, significant reduction in the number of queue rearrangements and fairer service as compared to STF. Performance is examined for various workloads. To our knowledge, such an analysis of epoch task scheduling has not appeared in the research literature for this type of system operating with our workload models.

The structure of this chapter is as follows: Section 5.2 specifies system and workload models, describes task routing and task scheduling strategies, presents the performance metrics and also describes model implementation and input parameters. In Sec. 5.3 experimental results and performance analysis are presented. The last section contains conclusions and suggestions for further research.

5.2. Model and Methodology

The technique used to evaluate the performance of the scheduling disciplines is experimentation using a synthetic workload simulation.

5.2.1. *System and Workload Models*

An open queuing network model of a distributed system is considered (Fig. 5.1). Jobs arrive to the system with mean arrival rate λ. $P = 16$ homogeneous processors are available, each serving its own queue.

Jobs consist of a set of $n \geq 1$ tasks that can be run in parallel. The number of tasks that a job consists of is this job's *degree of parallelism*. Each task is assigned to a processor queue according to the routing policy employed. Tasks are processed according to the current scheduling method. No migration or pre-emption is permitted. More than one tasks of the same job may be assigned to the same processor. On completing execution, a task waits at the join point for sibling tasks of the same job to complete

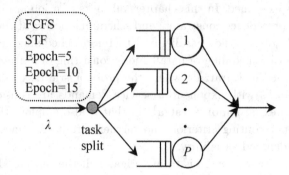

Fig. 5.1. The queuing network model. FCFS, STF, Epoch=5, Epoch=10, Epoch=15 are the scheduling methods, and λ is the job mean arrival rate.

execution. Therefore, task synchronisation is required which can seriously degrade parallel performance.

The workload considered here is characterised by three parameters: the distribution of job arrival, the distribution of the number of tasks per job, and the distribution of task service demand.

Distribution of Job Arrival. Job inter-arrival times are exponential random variables with a mean of $1/\lambda$.

Distribution of the Number of Tasks per Job. The number of tasks per job is uniformly distributed in the range of $[1..P]$. The mean number of tasks per job is equal to the $\eta = (1 + P)/2$.

Distribution of Task Service Demand. The impact of the variability in task service demand on system performance is examined. The following cases with regard to task service demand distribution are examined (μ represents the mean processor service rate): a) task service demands are exponentially distributed with a mean of $1/\mu$, and b) task service demands have a Branching Erlang distribution with two stages. The coefficient of variation is C, where $C > 1$ and the mean is $1/\mu$.

5.2.2. Task Routing Methods

Probabilistic Routing. With this method, a task is dispatched randomly to processors with equal probability.

Shortest Queue Routing. This policy assigns each ready task to the currently shortest processor queue.

5.2.3. Scheduling Strategies

First-Come-First-Served (FCFS). With this strategy, tasks are scheduled in a queue in the order of their arrival.

Shortest-Task-First (STF). This policy assumes that *a priori* knowledge about a task is available in form of service demand. When such knowledge is available, tasks in the processor queues are ordered in a decreasing order of service demand.

Epoch STF (ESTF) - Epoch = x. With this policy, processor queues are rearranged only at the end of epochs using the STF criterion. The size of an epoch is x.

5.2.4. Performance Metrics

Response time of a random job is defined as the interval of time from the dispatching of this job tasks to processor queues to service completion of the last task of this job. Parameters used in simulation computations are shown in Table 5.1. *RT* represents overall performance, while *MRT* provides an indication of fairness in individual job service.

5.2.5. Model Implementation and Input Parameters

The queuing network model described above is implemented with discrete event simulation using the independent replication method. For every mean value, a 95% confidence interval was evaluated. All confidence intervals were found to be less than 5% of the mean values.

Mean processor service time is chosen as $1/\mu = 1$. Since there are on average 8.5 tasks per parallel job $((P+1)/2)$, if all processors are busy, then an average of $P/8.5 = 1.88235$ parallel jobs can be served each unit of time. This implies that it must be chosen $\lambda < 1.88235$, and consequently $1/\lambda > 0.531$. For this reason it has been set $1/\lambda = 0.62, 0.60, 0.58, 0.56$, which corresponds to $\lambda = 1.613, 1.667, 1.724, 1.786$. Mean processor utilisation is given by the following formula:

$$U = \frac{\lambda \cdot \eta}{P \cdot \mu}$$

For the given values of μ and λ the expected mean processor utilisation is given in Table 5.2.

Epoch size is 5, 10, and 15. Epoch size 5 has been chosen as a starting point for the experiments because the mean processor service time is equal to 1, and also because with this epoch size *NQR* is significantly smaller

Table 5.1. Notations

Parameter	Definition
λ	Mean job arrival rate
$1/\mu$	Mean task service demand
η	Mean number of tasks per job
C	Coefficient of variation of task service demand
U	Mean processor Utilisation
RT	Mean Response Time
MRT	Maximum Response Time
NQR	Number of Queue Rearrangements
E	Estimation error in service time
D_{RT}	Relative (%) decrease in RT when the STF or the ESTF method is employed instead of FCFS
D_{NQR}	Relative (%) decrease in NQR when ESTF scheduling is employed instead of the STF policy
MRT Ratio	The ratio of MRT in the STF or ESTF scheduling case over MRT in the FCFS case

Table 5.2. Mean processor utilisation.

$1/\lambda$	U
0.62	0.8569
0.60	0.8856
0.58	0.9106
0.56	0.9488

than in the STF case. Therefore it was expected that larger epoch sizes would result in even smaller NQR.

Experiments for $C = 1$, 2, and 4 have been conducted. Also, service time estimation error $E = \pm 10\%$ and $\pm 20\%$ has been examined.

5.3. Simulation Results and Performance Analysis

Results of the probabilistic and shortest queue routing are presented next. Simulation experiments accessed the impact of service time estimation error on the performance of the scheduling methods in both routing cases. The simulation results (not shown here) indicated that estimation error in processor service time slightly affected performance when $E = 10, 20$. This is in accordance with other results reported in the literature which are related to estimation of service time.[3,7] For this reason results only for $E = 0$ are included.

5.3.1. Probabilistic Routing

As it is shown in Figs. 5.2 and 5.3 the worst performance for $C = 1$ appears with the FCFS policy because this method yields the highest mean response time. It is obvious that with FCFS scheduling, tasks with small service demand get blocked behind a task that has a large service demand and is waiting in the queue. Blocking behind a large task introduces inordinate queuing delays and also synchronisation delay to the sibling tasks.

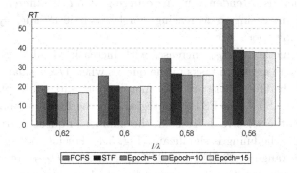

Fig. 5.2. RT versus $1/\lambda$, $E = 0$, $C = 1$, probabilistic routing.

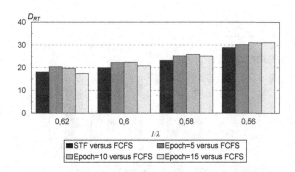

Fig. 5.3. D_{RT} versus $1/\lambda$, $E = 0$, $C = 1$, probabilistic routing.

It is also shown that mean response time with ESTF in some cases is lower than that of the STF method while in some other cases is very close to that of the STF method. This is due to the following reason. With the STF method tasks that have a small service demand do not delay behind a large task. However, this does not necessarily mean that the job to which they belong to will have a shorter response time. This is because there might be a case where some sibling tasks of the same job may have to delay longer in other queues due to the STF criterion. This may result in increased synchronisation delay of sibling tasks and therefore in increased response times of the respective jobs.

At each load RT does not differ significantly at different epoch sizes. For $1/\lambda = 0.62$ RT slightly increases with increasing epoch size, while as $1/\lambda$ decreases there is a tendency in the ordering of RT at different epochs to change from increasing to decreasing. For $1/\lambda = 0.56$ RT slightly decreases with increasing epoch size.

With all methods, D_{RT} increases with increasing load (Fig. 5.3). This is because there are more tasks in the queues when $1/\lambda$ is small than when is large and therefore there are then more opportunities to exploit the advantages of the STF and ESTF strategies over the FCFS method.

For all λ, the relative decrease in the number of queue rearrangements due to epoch scheduling is significant (Fig. 5.4). For all epoch sizes D_{NQR} varies in the range of 80.05% - 93.34%. For each λ, D_{NQR} increases with increasing epoch size. D_{NQR} increase is larger when epoch size is small than when is large. That is, in all cases the difference in D_{NQR} between epochs 5 and 10 is larger than the difference between epochs 10 and 15. For each epoch size, D_{NQR} is almost the same at different loads.

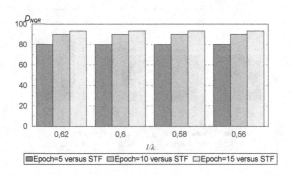

Fig. 5.4. D_{NQR} versus $1/\lambda$, $E = 0$, $C = 1$, probabilistic routing.

In Fig. 5.5 it is shown that STF and ESTF yield larger maximum response time than the FCFS method does. This is due to the fact that with the STF and ESTF policies, some tasks with large service demands suffer

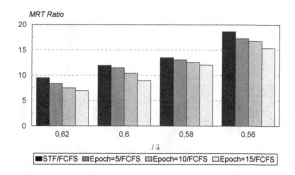

Fig. 5.5. *MRT Ratio* versus $1/\lambda$, $E = 0$, $C = 1$, probabilistic routing.

long delays in their respective queues and this results in large synchronisation delays of their sibling tasks and therefore in large response times of the respective jobs.

In all cases that are examined, ESTF is fairer than STF because the former method prevents tasks from unbounded delays. For any λ, the *MRT Ratio* decreases with increasing epoch size. Therefore, fairer service is provided to jobs when queue rearrangements take place after long epochs than after small ones.

From the results presented in Fig. 5.5 it is also shown that with each of the STF and ESTF methods, *MRT Ratio* increases with increasing load. This is due to the fact that when a task has a large service demand, then it is more probable to be bypassed by many tasks when the system load is high, then when the system load is low.

As shown in Figs. 5.6 and 5.7 the superiority of each of STF and ESTF over FCFS is more significant in the $C = 2$ case than in the $C = 1$. This is due to the fact that as C increases the variability in task service demand increases too and this results in better exploitation of the STF and ESTF strategies abilities. Also the results reveal that STF performs better than ESTF and that RT increases with increasing epoch size. In all cases ESTF for epoch size 5 performs close to STF. As in the $C = 1$ case, D_{RT} also increases with increasing load.

For all λ, simulation results (not shown here) reveal that D_{NQR} is almost the same as in the $C = 1$ case.

In Fig. 5.8 it is shown that the observations about *MRT Ratio* that hold for the $C = 1$ case also hold for $C = 2$. Furthermore, by comparing the results presented in Figs. 5.5 and 5.8 it is apparent that at each load

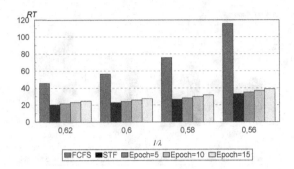

Fig. 5.6. RT versus $1/\lambda$, $E = 0$, C = 2, probabilistic routing.

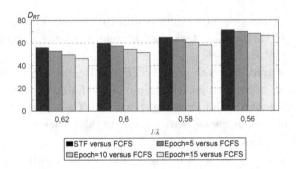

Fig. 5.7. D_{RT} versus $1/\lambda$, $E = 0$, C = 2, probabilistic routing.

MRT Ratio is larger in the $C = 1$ case than in the $C = 2$. It is also shown that for each epoch size the difference in *MRT Ratio* between STF and ESTF is smaller in the $C = 2$ case than in the $C = 1$.

As shown in Figs. 5.9 and 5.10 the superiority of each of STF and ESTF over FCFS is more significant in the $C = 4$ case than in the $C = 2$, due to the larger variability in task service demand in the former case. Also the results reveal that STF performs better than ESTF and that RT increases with increasing epoch size. ESTF for epoch size 5 performs close to STF. As in the other C cases D_{RT} increases with increasing load.

For all λ, D_{NQR} is almost the same as in the $C = 1$ and $C = 2$ cases.

In Fig. 5.11 it is shown that for each load all scheduling methods yield almost the same *MRT Ratio*. Therefore, for $C = 4$ STF and ESTF do not differ in fairness. The increase in *MRT Ratio* due to increasing load is less significant than in the $C = 1$ and $C = 2$ cases.

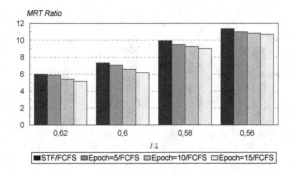

Fig. 5.8. *MRT Ratio* versus $1/\lambda$, $E = 0$, C = 2, probabilistic routing.

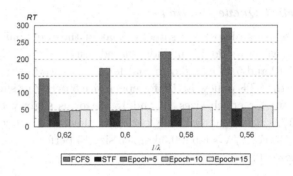

Fig. 5.9. *RT* versus $1/\lambda$, $E = 0$, $C = 4$, probabilistic routing.

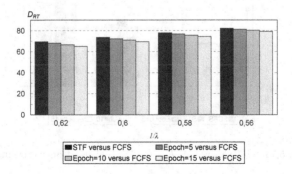

Fig. 5.10. D_{RT} versus $1/\lambda$, $E = 0$, $C = 4$, probabilistic routing.

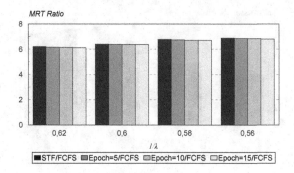

Fig. 5.11. *MRT Ratio* versus $1/\lambda$, $E = 0$, $C = 4$, probabilistic routing.

5.3.2. Shortest Queue Routing

For all C, D_{NQR} is almost the same as that of the probabilistic routing case, for this reason we do not include related results.

In Figs. 5.12 and 5.13 it is shown that for $C = 1$ the worst performance appears with the STF policy. ESTF for epoch size 5 performs slightly worse than FCFS but for epoch sizes 10 and 15 performs slightly better than FCFS. For $1/\lambda = 0.62, 0.60$ ESTF performs almost the same for epoch sizes 10 and 15. For $1/\lambda = 0.58, 0.56$ epoch size 15 performs slightly better than epoch size 10.

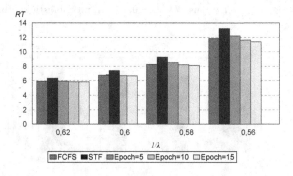

Fig. 5.12. *RT* versus $1/\lambda$, $E = 0$, $C = 1$, shortest queue routing.

There is the following explanation for the different relative performance of the scheduling methods under the probabilistic and the shortest queue routing. In the probabilistic routing case when the FCFS scheduling method

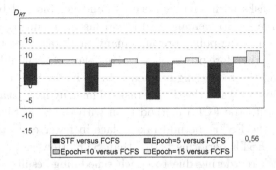

Fig. 5.13. D_{RT} versus $1/\lambda$, $E = 0$, $C = 1$, shortest queue routing.

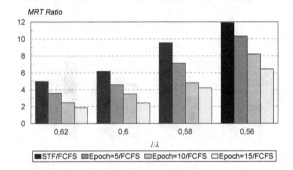

Fig. 5.14. *MRT Ratio* versus $1/\lambda$, $E = 0$, $C = 1$, shortest queue routing.

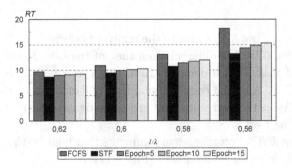

Fig. 5.15. RT versus $1/\lambda$, $E = 0$, $C = 2$, shortest queue routing.

is employed, tasks with small service demand get blocked behind a task that has a large service demand and is waiting in the queue. However, the shortest queue routing policy prevents many small tasks getting blocked in that queue and therefore prevents excessive delays of their sibling tasks. On the other hand the reordering of queued tasks in the STF case may cause synchronisation delays to the sibling tasks. This may have as a result better performance with the FCFS method than with the STF. Since reordering of tasks with the ESTF method takes place in fewer cases than with the STF method, ESTF performs better than STF. However, for epoch sizes 10 and 15 task reordering due to epoch scheduling results in some small performance improvement as compared to the FCFS case.

In Fig. 5.14 it is shown that regarding maximum response time the same comments that hold for the probabilistic case also hold here.

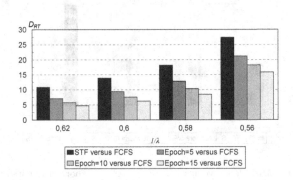

Fig. 5.16. D_{RT} versus $1/\lambda$, $E = 0$, $C = 2$, shortest queue routing.

As shown in Figs. 5.15 and 5.16 the relative performance of the scheduling policies for $C = 2$ is different than that of the $C = 1$ case due to the variability of task service demand. Here the worst method is FCFS and the best is STF. Also the results reveal that RT increases with increasing epoch size. As in the corresponding probabilistic case D_{RT} also increases with increasing load.

Regarding *MRT Ratio* the same comments that hold for the corresponding probabilistic case also hold here (Fig. 5.17).

As shown in Figs. 5.18 and 5.19 all comments about RT and D_{RT} that hold for $C = 2$ also hold for $C = 4$. Also the results reveal that the superiority of each of STF and ESTF over FCFS is more significant in the $C = 4$ case than in the $C = 2$ case.

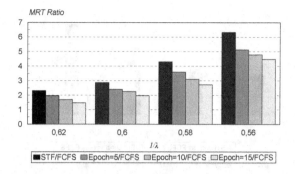

Fig. 5.17. *MRT Ratio* versus $1/\lambda$, $E = 0$, $C = 2$, shortest queue routing.

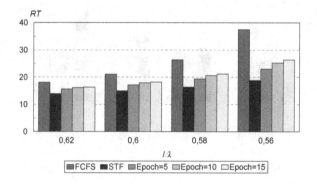

Fig. 5.18. *RT* versus $1/\lambda$, $E = 0$, $C = 4$, shortest queue routing.

In Fig. 5.20 it is shown that *MRT Ratio* is smaller here than in the $C = 2$ case. Therefore regarding fairness FCFS differs from STF and ESTF in a smaller degree here than in the $C = 2$ case. Also STF and ESTF for all epoch sizes differ between each other in fairness in a smaller degree when $C = 4$ than when $C = 2$. For $1/\lambda = 0.62$ *MRT Ratio* is almost the same in all methods cases. For $1/\lambda > 0.62$ *MRT Ratio* in the ESTF cases is smaller than in the STF case. Furthermore *MRT Ratio* generally slightly decreases with increasing epoch size. As in the other cases *MRT Ratio* increases with increasing load. The increase is less significant here than in the $C = 1$ and $C = 2$ cases.

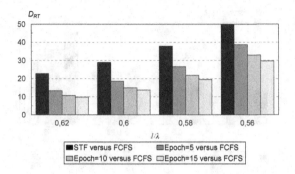

Fig. 5.19. D_{RT} versus $1/\lambda$, $E = 0$, $C = 4$, shortest queue routing.

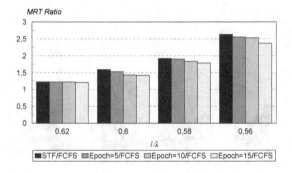

Fig. 5.20. *MRT Ratio* versus $1/\lambda$, $E = 0$, $C = 4$, shortest queue routing.

5.4. Conclusions

Distributed computing promises to meet increasing computational needs required in scientific research. A range of applications consisting of independent tasks can be assigned to distributed processors and run in parallel. Efficient task scheduling is very important because inappropriate scheduling may not exploit the potential of a distributed system limiting the gains of parallelisation.

STF is often used as a benchmark in scheduling studies. By using simulation over a spectrum of workload parameters, this article is able to evaluate the advantages and disadvantages of an alternative to the STF method, called epoch STF (ESTF), in comparison to traditional STF.

The simulation results reveal that the relative performance of STF and ESTF depends on the workload. In some cases ESTF (for some epoch sizes) performs close to or even better than STF. In all cases, epoch scheduling requires significantly fewer queue rearrangements (overhead) than STF. Large epochs involve smaller overhead than short epochs. Also, in most cases epoch scheduling is fairer than STF, while in the remaining cases the two methods do not differ significantly.

However, when exponential distribution is employed, the shortest queue routing simulation reveals that neither STF nor ESTF should be used. This is because STF performs worse than FCFS and ESTF performs either slightly worse or slightly better than FCFS.

Therefore, there are cases where ESTF is a better option for distributed systems than conventional STF scheduling. It is also interesting to compare "epoch" versions of other scheduling policies with the original methods.

This chapter presents cases where the number of tasks per job is bounded by the number of distributed processors in the system. It is interesting to consider cases where the number of tasks per job can be larger than the number of processors.

References

1. Aguilar J., Gelenbe E., "Task Assignment and Transaction Clustering Heuristics for Distributed Systems" *Information Sciences*, Elsevier Science, Amsterdam, Netherlands, **97**(2), (1997), 199-219.
2. Dail H., Berman F. and Casanova H., "A Decoupled Scheduling Approach for Grid Application Development Environments" *J. Parallel Distrib. Comput.*, Academic Press, Amsterdam, The Netherlands, **63**, (2003), 505-524.
3. Dandamudi S., "A Comparison of Task Scheduling Strategies for Multiprocessor Systems" IEEE Computer Society, Los Alamitos, CA, 1991, 423-426. (Proc. of the IEEE Symposium on Parallel and Distributed Processing, 1991)
4. Dandamudi S., "Performance Implications of Task Routing and Task Scheduling Strategies for Multiprocessor Systems" IEEE Computer Society, Los Alamitos, CA, 1994, 348-353. (Proc. of the IEEE-Euromicro Conference on Massively Parallel Computing Systems, 1994)
5. Dandamudi S., *Hierarchical Scheduling in Parallel and Cluster Systems* (1st Ed. Kluwer Academic/Plenum Publishers, New York) 2003.
6. Frachtenberg E., Feitelson D. G., Petrini F. and Fernandez J., "Flexible CoScheduling: Mitigating Load Imbalance and Improving Utilization of Heterogeneous Resources" IEEE Computer Society, Los Alamitos, CA, 2003, 85b. (Proc. of the International Parallel and Distributed Processing Symposium, 2003) (full paper in IEEE Computer Society Digital Library)
7. Karatza H. D., "Simulation Study of Task Scheduling and Resequencing in a Multiprocessing System" *Simulation Journal*, SCS, San Diego, CA, **68**(4), (1997), 241-247.

8. Karatza H. D., "Epoch Scheduling in a Distributed System" Eurosim, Delft, Netherlands, 2001, 1-6. (Proc. of the Eurosim 2001 Congress, 2001)
9. Karatza H. D., "Epoch Task Cluster Scheduling in a Distributed System" SCS, San Diego, CA, 2002, 259-265. (Proc. of the 2002 International Symposium on Performance Evaluation of Computer and Telecommunication Systems, 2002)
10. Karatza H. D., "A Comparison of Load Sharing and Job Scheduling in a Network of Workstations" *Intern. J. of Simulation: Systems, Science & Technology*, UK Simulation Society, Nottingham, UK, 4(3&4), (2003), 4-11.
11. Karatza H. D., "Epoch Task Scheduling in Distributed Server Systems" SCS Europe, Erlangen, Germany, 2004, 103-108. (Proc. of the 18th European Simulation Multiconference, 2004)
12. Kwok Y.-K., "On Exploiting Heterogeneity for Cluster Based Parallel Multithreading Using Task Duplication" *The J. of Supercomputing*, Kluwer Academic Publishers, The Netherlands, Amsterdam, **25**, (2003), 63-72.
13. Legrand A., Marchal L. and Casanova H., "Scheduling Distributed Applications: the SimGrid Simulation Framework" IEEE Computer Society, Los Alamitos, CA, 2003, 145-152. (Proc. of the 3rd IEEE/ACM International Symposium on Cluster Computing and the Grid, 2003)
14. McCann C. and Zahorjan J., "Scheduling Memory Constrained Jobs on Distributed Memory Parallel Computers" The Association for Computing Machinery, New York, USA, 1995, 208-219. (Proc. of the 1995 ACM Sigmetrics Conference, 1995)
15. Nikolopoulos D. S. and Polychronopoulos C. D., "Adaptive Scheduling Under Memory Constraints on Non-Dedicated Computational Farms" *Future Generation Computer Systems* Elsevier, Amsterdam, **19**, (2003), 505-519.
16. Sabin G., Kettimuthu R., Rajan A. and Sadayappan P., "Scheduling of Parallel Jobs in a Heterogeneous Multi-Site Environment" in "Job Scheduling Strategies for Parallel Processing", LNCS 2862 – Springer, 2003, 87-104.
17. Weissman J. B., Abburi L. R. and England D., "Integrated Scheduling: the Best of Both Worlds" *J. Parallel Distrib. Comput.*, Elsevier Science, New York, USA, **63**, (2003), 649-668.
18. Zhang Y., Franke H., Moreira J. and Sivasubramaniam A., "The Impact of Migration on Parallel Job Scheduling for Distributed Systems" LNCS 1900 – Springer, 2000, 242-251. (Proc. of Europar, 2000)

New Challenges on Modelling and Simulation

CHAPTER 6

Counter Intuitive Aspects of Statistical Independence in Steady State Distributions

Jeffrey P. Buzen

12 Mountain Laurels
Nashua, NH 03062 USA
E-mail: jeffbuzen@comcast.net

Stochastic processes that differ materially from one another can sometimes possess exactly the same steady state distribution. This makes it dangerous to draw certain inferences about the nature of a stochastic process by analysing the properties of its associated steady state distribution. Product form queuing networks provide a concrete example of this phenomenon by demonstrating that statistical independence among the states of individual queues does not necessarily imply that these queues operate independently in the associated stochastic process. The distinction between statistical independence and dynamic independence is introduced clarify the issues involved.

6.1. Introduction

The observed behaviour of real world systems often exhibits noteworthy degrees of variability. For example, the time between successive arrivals of customers at a queue is seldom the same from one customer to the next. Similarly, the time required to complete individual service requests often differs significantly among customers.

The most common approach for representing this type of variability is to assume that the real world system being studied can be modelled mathematically by a stochastic process. If an appropriate model can be formulated, the behaviour sequences (sample paths) that are associated with the stochastic process will exhibit the same type of variability that the system itself exhibits.

Once this initial objective has been achieved, the stochastic process can then be used to analyse, understand and predict the performance of the original real world system. Analysts pursuing this basic approach typically begin by deriving the steady state distribution of the underlying stochastic process. Mathematical expressions that characterise the behaviour of the system itself are then expressed as functions of this distribution.

Steady state stochastic models have been employed successfully for many decades in a number of branches of science and engineering. As a result, many analysts have become highly proficient in the required mathematical techniques. However, applying these techniques without fully understanding all assumptions implicit in the analysis can sometimes lead to confusing and erroneous conclusions. This discussion will explore some of the less well understood assumptions used in stochastic modelling with the goal of shedding light on areas of potential misunderstanding.

6.2. A System of Two Independent M/M/1 Queues

The main points in this paper can be illustrated through a few simple examples. Begin by considering Fig. 6.1, which represents two completely independent M/M/1 queues. Following the standard conventions of queuing theory, assume that arrivals at Queue 1 are generated by a Poisson process with mean rate λ_1. Assume also that Queue 1 is served by a single server whose service times are characterised by a sequence of independent exponentially distributed random variables with mean $1/\mu_1$.

Queue 2 is characterised in a similar manner. Arrivals are assumed to be generated by a Poisson process with the same mean rate λ_1, while service times for the single server at Queue 2 are characterised by a sequence of independent exponentially distributed random variables with mean $1/\mu_2$.

The steady state distribution of an individual M/M/1 queue is perhaps the most widely studied result in queuing theory. Extending this result to the case of the two independent M/M/1 queues shown in Fig. 6.1 is straightforward. The details are presented here to provide a point of reference for the subsequent discussion.

Begin by representing the state of this system as (n_1, n_2), where n_1 corresponds to the number of customers at Server 1 and n_2 corresponds to the number of customers at Server 2. Thus, n_1 and n_2 both range over the set of non-negative integers: $\{0, 1, 2, ... \}$.

Let $P(n_1, n_2)$ denote the steady state distribution for this system of queues. That is, $P(n_1, n_2)$ is the steady state probability that there are n_1 customers at Server 1 and n_2 customers at Server 2.

The key to deriving an expression for $P(n_1, n_2)$ is to write down a set

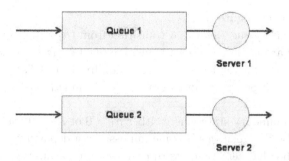

Fig. 6.1. Two independent M/M/1 queues.

of equations representing the fact that, in steady state, the rate of transition into each system state is equal to the rate of transition out. Only transitions that occur as a result of single arrivals or departures need be considered. Transitions that result from simultaneous arrivals and departures of multiple customers are ignored since such simultaneous events occur with probability zero.

Under these assumptions, there are exactly four transitions into each "non-boundary" state where $n_1 > 0$ and $n_2 > 0$. There are also exactly four transitions out. The eight transitions into and out of each non-boundary state are represented by arrows in Fig. 6.2 and specified as follows:

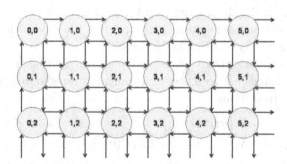

Fig. 6.2. State transition diagram for two independent M/M/1 queues.

Transitions out of state (n_1, n_2):
1. An arrival at Queue 1 causes a transition from (n_1, n_2) to $(n_1 + 1, n_2)$.
2. An arrival at Queue 2 causes a transition from (n_1, n_2) to $(n_1, n_2 + 1)$.
3. A departure at Server 1 causes a transition from (n_1, n_2) to $(n_1 - 1, n_2)$.
4. A departure at Server 2 causes a transition from (n_1, n_2) to $(n_1, n_2 - 1)$.

Transitions into state (n_1, n_2):
5. An arrival at Queue 1 causes a transition from $(n_1 - 1, n_2)$ to (n_1, n_2).
6. An arrival at Queue 2 causes a transition from $(n_1, n_2 - 1)$ to (n_1, n_2).
7. A departure at Server 1 causes a transition from $(n_1 + 1, n_2)$ to (n_1, n_2).
8. A departure at Server 2 causes a transition from $(n_1, n_2 + 1)$ to (n_1, n_2).

For each boundary state where either $n_1 = 0$ or $n_2 = 0$ (but not both), two of the above transitions become impossible: a departure cannot occur at a server that has zero requests to process (eliminating the possibility of Transitions 3 or 4), and an arrival cannot cause queue length to increase from minus one to zero since queues of length minus one cannot exist (eliminating the possibility of Transitions 5 or 6). The top row and left column of Fig. 6.2 illustrate the six remaining transitions for each of these boundary states.

Finally, in the special origin state (0,0), both servers are idle and both queues are empty. This eliminates the possibility of Transitions 3, 4, 5 and 6, leaving only the four transitions shown in the upper left corner of Fig. 6.2.

The next step in deriving the steady state distribution for $P(n_1, n_2)$ is to write down the set of equations expressing the fact that, in steady state, the aggregate rate of transition into each state is equal to the aggregate rate out. To calculate individual transition rates, multiply the proportion of time spent in a state by the rate at which the corresponding transition occurs. For example, since arrivals to Queue 1 occur at rate λ_1, the rate at which transitions from state (n_1, n_2) to $(n_1 + 1, n_2)$ occur as a result of arrivals to Queue 1 is equal to $\lambda_1 P(n_1, n_2)$.

For non-boundary states where $n_1 > 0$ and $n_2 > 0$, each equation expressing a balance between the total transition rates in and out involves eight separate transitions:

$[\lambda_1 + \lambda_1 + \mu_1 + \mu_2]P(n_1, n_2) =$
$\lambda_1 P(n_1 - 1, n_2) + \lambda_1 P(n_1, n_2 - 1) + \mu_1 P(n_1 + 1, n_2) + \mu_2 P(n_1, n_2 + 1)$

For the boundary states where either n_1 or n_2 equals 0 (but not both), each equation involves six separate transitions:

$[\lambda_1 + \lambda_1 + \mu_1]P(n_1, 0) = \lambda_1 P(n_1 - 1, 0) + \mu_1 P(n_1 + 1, 0) + \mu_2 P(n_1, 1)$
$[\lambda_1 + \lambda_1 + \mu_2]P(0, n_2) = \lambda_1 P(0, n_2 - 1) + \mu_1 P(1, n_2) + \mu_2 P(0, n_2 + 1)$

Finally, only four transitions are involved in balancing the aggregate flow into and out of the origin state (0,0):

$$[\lambda_1 + \lambda_1]P(0,0) = \mu_1 P(1,0) + \mu_2 P(0,1)$$

Assuming $\lambda_1 < \mu_1$ and $\lambda_1 < \mu_2$, it is easy to show that this set of linear equations has the following product form solution.

$$P(n_1, n_2) = N_1 N_2 (\lambda_1/\mu_1)^{n_1} (\lambda_1/\mu_2)^{n_2} \qquad (1)$$

where $N_1 = (1 - \lambda_1/\mu_1)$ and $N_2 = (1 - \lambda_1/\mu_2)$. The solution can be verified by simply substituting it into the balance equations on the previous page. Note that N_1 and N_2 have no impact on the balance equations. Instead, these two factors are normalising constants, chosen to insure the sum of $P(n_1, n_2)$ over all non-negative values of n_1 and n_2 is equal to 1.

The fact that $P(n_1, n_2)$ has this particular form should come as no surprise. Since the two queues in Fig. 6.1 are independent, one would expect $P(n_1, n_2)$ to be the product of $P(n_1)$ and $P(n_2)$, where $P(n_1)$ and $P(n_2)$ are the steady state distributions associated with Queue 1 and Queue 2 respectively. In fact, this assumption is easy to demonstrate since the well known solution to the M/M/1 queue yields:

$$P(n_1) = N_1 (\lambda_1/\mu_1)^{n_1} \qquad (2)$$
$$P(n_2) = N_2 (\lambda_1/\mu_2)^{n_2} \qquad (3)$$

Thus,

$$P(n_1, n_2) = P(n_1) P(n_2) \qquad (4)$$

6.3. A System of Two Queues in Tandem

Now consider the system of queues shown in Fig. 6.3. Queue 1 has exactly the same properties as Queue 1 in Fig. 6.1. Arrivals are generated by a Poisson process with mean rate λ_1, and service times are characterised by a sequence of independent exponentially distributed random variables with mean $1/\mu_1$.

In the case of Queue 2, service times are again characterised by a sequence of independent exponentially distributed random variables with mean $1/\mu_2$. However, the arrival process for Queue 2 in Fig. 6.3 is specified in a new manner. Instead of being characterised by a Poisson arrival process with rate λ_1, it is assumed that arrivals at Queue 2 are generated by departures from Queue 1. Thus, if the state of these two queues is again represented as (n_1, n_2), each time n_1 is reduced by 1 (due to a departure at Queue 1), there is a simultaneous increase of plus 1 in the value of n_2. It is clear that the two queues are tightly linked and do not operate independently.

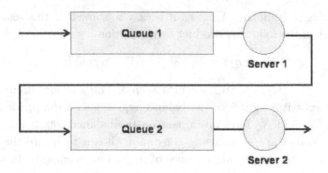

Fig. 6.3. Two queues in tandem.

As before, the key to deriving an expression for $P(n_1, n_2)$ is to write down a set of equations representing the fact that, in steady state, the rate of transition into state (n_1, n_2) is equal to the rate of transition out. Applying the assumptions used in the previous section, there are now exactly three transitions into each "non-boundary" state where $n_1 > 0$ and $n_2 > 0$. There are also exactly three transitions out. The six transitions into and out of each non-boundary state are represented by arrows in Fig. 6.4 and specified as follows:

Transitions out of state (n_1, n_2):
1. An arrival at Queue 1 causes a transition from (n_1, n_2) to $(n_1 + 1, n_2)$.
2. A departure at Server 1 causes a transition from (n_1, n_2) to $(n_1 - 1, n_2 + 1)$.
3. A departure at Server 2 causes a transition from (n_1, n_2) to $(n_1, n_2 - 1)$.

Transitions into state (n_1, n_2):
4. An arrival at Queue 1 causes a transition from $(n_1 - 1, n_2)$ to (n_1, n_2).
5. A departure at Server 1 causes a transition from $(n_1 + 1, n_2 - 1)$ to (n_1, n_2).
6. A departure at Server 2 causes a transition from $(n_1, n_2 + 1)$ to (n_1, n_2).

Note that Transitions 2 and 5 reflect the tandem coupling between the two queues. These transitions correspond to the diagonal arrows in Fig. 6.4.

As in the previous example, two of these transitions become impossible in those boundary states where either $n_1 = 0$ or $n_2 = 0$ (but not both): a departure cannot occur at a server that has zero requests to process (eliminating the possibility of Transitions 2 or 3), and an arrival cannot cause queue length to increase from minus one to zero since queues of length minus one cannot exist (eliminating the possibility of Transitions 4

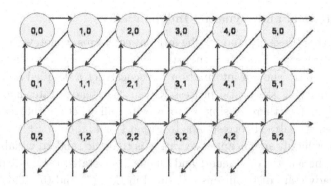

Fig. 6.4. State transition diagram for two queues in tandem.

or 5). The top row and left column of Fig. 6.4 illustrate the four remaining transitions for each of these boundary states.

Finally, in the special origin state (0,0), both servers are idle and both queues are empty, leaving only the two transitions shown in the upper left corner of Fig. 6.4.

Once again, the next step in deriving the steady state distribution for $P(n_1, n_2)$ is to write down the set of equations specifying that the transition rate into each state is equal to the transition rate out. For non-boundary states where $n_1 > 0$ and $n_2 > 0$, each equation involves six separate transitions:

$$[\lambda_1 + \mu_1 + \mu_2] P(n_1, n_2) = \lambda_1 P(n_1-1, n_2) + \mu_1 P(n_1+1, n_2-1) + \mu_2 P(n_1, n_2+1)$$

For the boundary states where either n_1 or n_2 equals 0 (but not both), each equation involves four separate transitions:

$$[\lambda_1 + \mu_1] P(n_1, 0) = \lambda_1 P(n_1 - 1, 0) + \mu_2 P(n_1, 1)$$
$$[\lambda_1 + \mu_2] P(0, n_2) = \mu_1 P(1, n_2 - 1) + \mu_2 P(0, n_2 + 1)$$

Finally, only two transitions are involved in balancing the aggregate flow into and out of the origin state (0,0):

$$\lambda_1 P(0, 0) = \mu_2 P(0, 1)$$

Assuming $\lambda_1 < \mu_1$ and $\lambda_1 < \mu_2$, it is easy to verify that (5) represents the solution to this set of linear equations. Note that (5) is exactly the same as (1), even though the systems shown in Figs. 6.1 and 6.3 are modelled by different stochastic processes.

$$P(n_1, n_2) = N_1 N_2 (\lambda_1/\mu_1)^{n_1} (\lambda_1/\mu_2)^{n_2} \qquad (5)$$

6.4. Statistical and Dynamic Independence

The fact that Eqs. (1) and (5) have identical forms is noteworthy for a number of reasons. First, it demonstrates that a steady state distribution does not contain enough information to reconstruct a detailed description of the underlying stochastic process. Given the steady state distribution presented in (5), there is simply no way to tell whether the real world system being analysed corresponds to Fig. 6.1 or Fig. 6.3.

Another highly significant observation is that the random variables representing the lengths of Queue 1 and Queue 2 are statistically independent in the steady state distributions presented in Eqs. (1) and (5). This follows immediately from (4), which states that the joint distribution $P(n_1, n_2)$ is the product of $P(n_1)$, the unconditional distribution of n_1, and $P(n_2)$, the unconditional distribution of n_2.

The fact that random variables n_1 and n_2 are statistically independent is not surprising in the case of Fig. 6.1 since queues 1 and 2 operate independently in this example. However, queues 1 and 2 do not operate independently in Fig. 6.3. They are linked by the fact that a service completion at Queue 1 triggers an immediate arrival at Queue 2. Given this linkage, it may be difficult to understand why random variables n_1 and n_2 are statistically independent in the corresponding steady state distribution.

This counter intuitive observation can, of course, be explained in purely algebraic terms by considering the balance equations. It has already been shown that the product form solution in (5) satisfies the balance equations for each node in the state transition diagram of Fig. 6.4. Since Fig. 6.4 is the state transition diagram for the system illustrated in Fig. 6.3, the random variables n_1 and n_2 that are associated with Fig. 6.3 must satisfy (5) and thus must be statistically independent.

To develop a more intuitive understanding of this surprising result, begin by assuming that the system in Fig. 6.3 is in steady state and is observed for some interval of time. If the observer happens to know the number of customers at Server 1 at any given instant, it is reasonable to ask whether this knowledge provides the observer with any additional information about the corresponding number of customers at Server 2.

The fact that n_1 and n_2 are statistically independent implies that this is not the case. Knowing the number of customers at Server 1 at any particular instant does not provide any additional information about the number of customers at Server 2. This tightly focused observation, which represents the essence of statistical independence, is entirely compatible with the fact that queues 1 and 2 are linked to one another as shown in Fig. 6.3.

Even though the obvious dependency that exists between Queue 1 and Queue 2 in Fig. 6.3 cannot be detected by studying the steady state distribution in (5), the relationship is displayed clearly in the state transition diagram of Fig. 6.4. The diagonal lines in the diagram represent transitions that occur when customers complete service at Queue 1 and proceed immediately to Queue 2. Queues that operate in a completely independent manner never have such diagonal lines in their state transition diagrams. For example, the state transition diagram in Fig. 6.2 contains only horizontal and vertical transitions, which is as expected since the two queues in Fig. 6.1 really do operate independently.

Another implication of the complete independence that exists between the two queues in Fig. 6.1 is that arrival rates at each queue are independent of the number of customers at the other queue. Specifically, the rate of transition from (n_1, n_2) to $(n_1 + 1, n_2)$ is independent of n_2, and the rate of transition from (n_1, n_2) to $(n_1, n_2 + 1)$ is independent of n_1.

In addition, completion rates (reciprocals of service times) at each queue are also independent of the number of customers at the other queue. Thus, the rate of transition from (n_1, n_2) to $(n_1 - 1, n_2)$ is independent of n_2, and the rate of transition from (n_1, n_2) to $(n_1, n_2 - 1)$ is independent of n_1.

These assumptions regarding transition rates, coupled with the assumption that there are no diagonal lines in the state transition diagram, imply that each queue operates in manner that is truly independent of the other. To distinguish this "true independence" from the more restrictive notion of statistical independence, it would seem appropriate to introduce a new term, *dynamic independence*, to denote this concept. Note that dynamic independence is a property of stochastic processes, whereas statistical independence is a property of random variables.

Figures 6.1 and 6.2, together with (1), illustrate that dynamic independence in a stochastic process implies statistical independence in the associated steady state distribution. However, Figs. 6.3 and 6.4, together with (5) illustrate that the converse is not true: statistical independence in the steady state distribution does necessarily imply dynamic independence in the associated stochastic process.

Although this discussion has been based on the simple queuing networks shown in Figs. 6.1 and 6.3, the points raised here actually apply to very broad class of queuing networks. In particular, Jackson[8] has shown that product form solutions can be derived for networks of the type shown in Fig. 6.2 that contain an arbitrary number of servers interconnected through an arbitrary set of linkages. Jackson's results have been generalised further by other researchers, notably Baskett, Chandy, Muntz and Palacios[1] and Gelenbe[6,12–15]. Replacing Fig. 6.3 by one of these more complex product form queuing networks would leave the main points of this discussion entirely unchanged.

6.5. Beyond Stochastic Modelling

Even though this discussion has focused exclusively on stochastic modelling, the results have implications for a broad range of models and analysis procedures. For example, the two systems of queues depicted in Figs. 6.1 and 6.3 can be analysed using a number of alternative approaches. In addition to a purely mathematical approach based on stochastic models, these systems can also be analysed by running computer simulation programs, by measuring performance within specially constructed experimental test beds [7,9,17,18], or by going out into the field and measuring the performance of actual systems operating under real word conditions.

If the assumptions of Poisson arrivals and exponentially distributed service times were incorporated into the simulations programs, the experimental test beds and the real world systems being measured, the two main conclusions discussed in the previous section would remain the same: measured or computed steady state distributions would be essentially identical for both Fig. 6.1 and Fig. 6.3, and this steady state distribution would have the product form, implying that queue lengths are independently distributed in both networks.

The fact that these main conclusions have been demonstrated in the context of a stochastic model provides certain immediate benefits. Using an analytic model eliminates concerns about measurement accuracy, and about transient effects associated with the duration of the simulation or measurement interval. However, the two main conclusions cited in the previous paragraph do not really depend upon the fact that stochastic models were used to analyse the systems in Figs. 6.1 and 6.3. They are equally valid under several alternative modelling and analysis procedures, assuming of course that the systems illustrated in Figs. 6.1 and 6.3 are the same for all analyses.

6.5.1. *Central Role of Steady State Distributions*

The key factor that leads all these procedures to a common set of conclusions is the use of steady state distributions to characterise the performance of complex systems. While the use of these distributions is both intuitively appealing and mathematically sound, it also raises a set of subtle issues that are worthy of further discussion.

For analysts working with real systems or experimental test beds, the collection and analysis of performance measurements leads naturally to the construction of steady state distributions. The first step in measuring the performance of any system operating over time is to specify the beginning and ending points of the measurement interval. Instruments (or software monitors) then continuously record the "state" of the system (e.g., the lengths of Queue 1 and Queue 2) throughout the interval. After the measurement interval has elapsed, it is reasonable to summarise the collected measurement data by dividing the total amount of time spent in each state by the length of the measurement interval. This results in a distribution that specifies the proportion of time the system spends in each state. Performance statistics such as mean queue length and response time can then be readily derived from this distribution.

This measurement-oriented steady state distribution, which can also be generated by running a computer simulation program, does not require the powerful mathematical constructs needed to define the steady state distribution associated with a stochastic process. From a mathematical perspective, a stochastic process is associated with a random variable (or a vector of random variables) whose distribution varies as a function of time. If this distribution converges to some stable distribution as time increases, and if this stable distribution is independent of the initial distribution at time zero, the stochastic process is said to be Ergodic. An Ergodic stochastic process whose distribution has stabilised (become independent of time) is said to be in equilibrium or in steady state. The resulting stable distribution is known as a steady state distribution.

Although measurement-oriented steady state distributions may seem very different from the mathematical steady state distributions associated with stochastic processes, they are tightly linked by an important result known as the Ergodic theorem. This theorem demonstrates that, if a Ergodic stochastic process in is steady state and is observed for a sufficiently long period of time, the measurement-oriented steady state distribution computed during the observation interval will converge (with probability one) to the underlying mathematical steady state distribution.

The basic mechanism that leads to the proof of the Ergodic theorem has a simple intuitive interpretation. It is well known that the empirical frequency distribution obtained by sampling any random variable for a large number of times will converge (with probability one) to the mathematical distribution associated with that random variable. Measuring the state of a stochastic process over a long observation interval is the continuous-time counterpart of constructing a frequency distribution by drawing a large (but discrete) number of samples from an ordinary random variable. If the stochastic process has reached steady state, the same random variable (i.e., the same distribution) will be sampled at each point in the continuous observation interval. Thus, the measurement-oriented steady state distribution will converge to the mathematical steady state distribution associated with the stochastic process.

As a result of this tight coupling between mathematical theory and empirical measurement procedures, the derivation and estimation of steady state distributions have become primary objectives in most analyses of system performance. This is true for analyses based on stochastic models, computer simulations, and measurements taken in experimental test beds and in real world operating environments.

6.5.2. Generality, Robustness and Level of Detail

Every model involves some form of abstraction that retains certain essential properties of the object being studied while discarding other details. By reducing highly detailed state-by-state measurement traces to a few numbers in an empirical distribution, steady state distributions can similarly be regarded as summaries that retain certain essential properties of an underlying system while eliminating others.

The notion of dynamic independence is a potentially important detail that is lost during the summarisation process followed when constructing a steady state distribution. As Fig. 6.1 and (1) illustrate, dynamic independence in the underlying system results in statistical independence in the steady state distribution. However, Fig. 6.3 and (5) illustrate that statistical independence in the steady state distribution can also arise even though there is no dynamic independence in the underlying system.

This means that researchers concerned about issues such as dynamic independence should not be content with analyses that yield only steady state distributions and derived quantities. Such researchers should remain alert to the fact that detecting subtle interactions among variables may require returning to the original measurement data and applying more powerful analytic techniques that can discover time dependent correlations that may exist even though a system has reached steady state. Detailed trace data generated by simulations and studies involving experimental test beds are especially well suited to such refined analyses.

On the other hand, it is also important to avoid being overly zealous in the extraction of highly detailed information from any analytic model, computer simulation or experiment conducted with a test bed or a real world system. To illustrate the essence of the problem, consider a simple variable such as the response time at a single queue. Since it is typical to assume that both service times and inter-arrival times are variable, one would not expect the response times of all customers to be identical.

The simplest way to characterise response time in this case is to report the mean. If more detail is desired, the standard deviation can also be computed. It is, of course, possible to go further by computing response time percentiles for any degree of granularity. At some point, these successively finer levels of detail can become essentially meaningless, reflecting highly specialised artifacts of the modelling assumptions or the measurement data. These details can seldom be generalised robustly to other cases. In addition, they are subject to dramatic change as a result of slight alterations in modelling assumptions.

In this context, metrics such as means and standard deviations that are derived directly from steady state distributions have the advantage of being relatively robust. This is due in part to the fact that the steady state distributions that these metrics depend upon are not sensitive to certain details of system behaviour such as the linkage between Queue 1 and Queue 2 that is illustrated in Fig. 6.3.

Thus, the fact that such linkages have no direct impact on the steady state distributions of the associated stochastic processes should not be construed as a negative observation. There is a natural tension between the level of detail captured in a solution and the generality and robustness of that solution. Following the principle embodied in Occam's razor, one can maximise generality and robustness by seeking a model with the least restrictive set of assumptions needed to derive the desired result. This principle has applications in the formulation of analytic models, the design of simulation programs and the construction of test beds for experimental studies.

6.5.3. *Operational Analysis*

One instructive illustration of this principle is the use of operational analysis to derive steady state distributions and related statistics without making all the assumptions required in a conventional stochastic model. Operational analysis is an alternative mathematical framework for formulating and deriving equations that characterise system performance. The conceptual basis for this framework is provided by the measurement-oriented discussion of steady state distributions presented earlier.

Operational analysis begins with the simple assumption that an analyst is measuring the behaviour of a real system or an experimental test bed during an interval of time. There is no need to assume that the real system or experimental test bed is related in any way to an underlying stochastic process. All that is necessary is to assume that measurements are being made during an observation interval, and that an analyst is interested in equations that relate these measurements to one another. Note that this direct approach avoids the need to invoke the Ergodic theorem when applying the resulting equations to measurement data collected under real world conditions and in experimental test beds.

It might seem that few, if any, interesting relationships among measured variables can be derived without any assumptions whatsoever. However, the first published paper on operational analysis[2] demonstrated that a number of results that are traditionally regarded as the cornerstones of stochastic queuing theory (Little's Law, the Utilisation Law, etc.) can in fact be derived under operational analysis with a minimum number of very simple assumptions. The robustness and generality of these results can thus be explained by the fact that they can be formally derived without appealing to the Ergodic theorem or the subtle statistical assumptions inherent in the concept of steady state stochastic processes.

Operational assumptions necessary to derive the steady state distribution of the M/M/1 queue (see (2)) were introduced almost immediately in a follow-on paper[3]. This paper demonstrated that the balance equations associated with the single queue version of Fig. 6.1 can be deduced from simple assumptions about observable system behaviour. In other words, the mathematical equations used to characterise the steady state distribution of an M/M/1 queue can be derived without requiring the assumption of an underlying stochastic process. These early results were subsequently extended to networks of queues[4] and to more general categories of models[5] and to stationary deterministic flows[10,11].

The fact that Eqs. (1) and (5) can be derived through operational analysis without appealing to any stochastic assumptions underscores a point made earlier: the inability of steady state distributions to distinguish between Fig. 6.1 and Fig. 6.3 reflects an inherent property of these distributions, and applies whether these distributions are measured experimentally or derived analytically within the formal mathematical frameworks provided by either stochastic modelling or operational analysis.

6.6. Conclusions

A steady state distribution characterises certain important properties of its associated stochastic process, but does not reflect all significant properties. Thus, two or more stochastic processes with materially different structures can have exactly the same steady state distribution. This fact limits the nature of the inferences that can be drawn regarding real world systems that are modelled as stochastic processes and analysed solely through steady state distributions.

In particular, the observation that two random variables are statistically independent in a steady state distribution does not automatically imply that the processes represented by these random variables operate independently in the real world. A new concept, dynamic independence, is introduced to characterise this type of independence. Dynamic independence is a property of a stochastic process rather than a steady state distribution.

Steady state distributions obtained by directly measuring performance in real world settings or experimental test beds, or by running simulation programs, exhibit the same property. Random variables associated with these measurement-oriented distributions can be statistically independent even though their counterparts are tightly linked to one another in the actual systems being investigated. Analysts interested in investigating such linkages need to go beyond the information contained in measurement-oriented steady state distributions. They may need to consider applying more powerful analytic techniques to the detailed traces that are capable of being generated by computer simulations and experimental test beds. The development of new analytic techniques for addressing these issues, especially techniques that do not require the complex assumptions of stochastic modelling but employ instead the less restrictive framework of operational analysis, represents a promising area for future research.

References

1. Baskett, F., Chandy, K.M., Muntz, R.R. and Palacios, J., "Open, Closed and Mixed Networks with Different Classes of Customers" *J. ACM.*, **22**(2), (April 1975), 248 - 260.
2. Buzen, J.P., "Fundamental Operational Laws of Computer Systems Performance" *Acta Informatica* **7**(2), (June 1976), 167-182.

3. Buzen, J.P., "Operational Analysis: The Key to the New Generation of Performance Prediction Tools" Washington, DC, 1976, 166-171. (Proc. IEEE COMPCON 76, Sep. 1976)
4. Denning, P.J. and Buzen, J.P., "Operational Analysis of Queuing Networks" in "Modelling and Performance Evaluation of Computer Systems" Beilner, H. and Gelenbe, E., eds. (North-Holland Publishing Company, Amsterdam) 1977, 151-172. (Proc. 3rd Int'l Symposium on Modeling and Performance Evaluation of Computer Systems, Oct. 1977)
5. Denning, P.J. and Buzen, J.P., "The Operational Analysis of Queuing Network Models" *ACM Computing Surveys*, **10**(3), (Sept. 1978), 225-261.
6. Gelenbe E. "G-networks with instantaneous customer movement", *Journal of Applied Probability*, **30**(3), (1993), 742-748.
7. Gelenbe, E., "Research and Development Test-Beds for Future Networks" Keynote Address. 2005, (1st Conf. on testbeds and research infrastructures for the development of network and communities, Trento, Italy, Feb. 2005)
8. Jackson, J.R., "Jobshop-like Queuing Systems" *Management Science*, **10**(1), (Oct. 1963), 131 - 142.
9. Newman, M.E., "The Structure and Function of Complex Networks" *SIAM Review*, **45**, (2003), 167-256.
10. Gelenbe E., "Stationary deterministic flows", *Theoretical Computer Science*, **23**,(1983), 107-128.
11. Gelenbe E., Finkel D., "Stationary deterministic flows II: the single server queue" *Theoretical Computer Science*, **52**, (1987), 269-280.
12. Gelenbe E. "Queueing networks with negative and positive customers", *Journal of Applied Probability*, **28**, (1991), 656-663.
13. Gelenbe E. "G-Networks with signals and batch removal", *Probability in the Engineering and Informatonal Sciences*, **7**, (1993), 335-342.
14. Gelenbe E., Fourneau J.M. "G-Networks with resets", *Performance Evaluation*, **49**, (2002), 179-192.
15. Fourneau J. M., Gelenbe E. "Flow equivalence and stochastic equivalence in G-Networks", *Computational Management Science*, **1**(2),(2004), 179 -192.
16. Gelenbe E."Learning in the recurrent random network", *Neural Computation*, **5**, (1993), 154-164.
17. Gelenbe E., Lent R., Xu Z., "Measurement and performance of a cognitive packet network", *Computer Networks*, **37**, (2001), 691-791.
18. Gelenbe E., Lent R., Nunez A. "Self-Aware networks and QoS", *Proceedings of the IEEE*, **92**(9), (2004), 1478-1489.

CHAPTER 7

The Nonstationary Loss Queue: A Survey

Khalid Abdulaziz Alnowibet

Department of Statistics and Operations Research, King Saud University
Riyadh 1145, Kingdom of Saudi Arabia
E-mail:knowibet@ksu.edu.sa

Harry Perros

Computer Science Department, NC State University
Raleigh, NC 27695-7534, USA
E-mail:hp@csc.ncsu.edu

The nonstationary loss queue is of great interest since the arrival rate in most communication systems varies over time. In view of the difficulty in solving the nonstationary loss queue, various approximation methods have been developed. In this paper, we review several of these approximation methods and present a new technique, the *fixed point approximation* (FPA) method. Numerical evidence points to the fact that the FPA method gives the exact solution.

7.1. Introduction

The loss queue is a queuing system consisting of s servers and no waiting room. A customer is lost if it arrives at a time when all servers are busy. The loss queue is commonly used to model the telephone network. It has also been recently used to model traffic-groomed optical networks and optical burst switching (OBS) networks, see for instance Washington and Perros[1] and Battestilli and Perros[2], and references within.

The loss queue has been extensively studied in the stationary case, i.e., assuming that the arrival rate and the service rate are time invariant. The nonstationary loss queue, where the arrival rate is time-dependent is also of great interest, since in most communication systems the arrival rate varies over time.

Consider a nonstationary loss queue $M(t)/M/s/s$ with a Poisson arrival process with a time-dependent rate $\lambda(t)$. Each arrival requests a service that requires an exponential amount of time with mean $1/\mu$. The service requested by an arriving customer is performed by a single server. The system has s identical servers, and there is no waiting room. The probability that there are n, $n = 0, 1, \ldots, s$, customers in the system at time t, $P_n(t)$, is represented by the following set of forward differential equations:

$$\frac{d}{dt}P_0(t) = \mu P_1(t) - \lambda(t)P_0(t),$$

$$\frac{d}{dt}P_n(t) = \lambda(t)P_{n-1}(t) + (n+1)\mu P_{n+1}(t) \\ - (\lambda(t) + n\mu)P_n(t), \quad 0 < n < s, \tag{1}$$

$$\frac{d}{dt}P_s(t) = \lambda(t)P_{s-1}(t) - s\mu P_s(t),$$

where $P_0(t) + P_1(t) + P_2(t) + \cdots + P_s(t) = 1$, $t \geq 0$, and $0 \leq P_n(t) \leq 1$ for $t \geq 0$ and $n = 0, 1, 2, \ldots, s$, with initial conditions: $P_0(0) = 1$ and $P_n(0) = 0$; $n = 1, 2, 3, \ldots, s$.

In the stationary case, where the arrival rate is constant, that is $\lambda(t) = \lambda$ for all t, we have that $(d/dt)P_n(t) \to 0$ for all n as $t \to \infty$. The above system of differential equations reduces to a set of linear equations from which we can obtain the familiar closed-form solution for the probability P_n that there are n customers in the system:

$$P_n(t) = \lim_{t \to \infty} P\{Q(t) = n\} = \frac{\rho^n/n!}{\sum_{i=0}^{s} \rho^i/i!}, \quad n = 0, 1, 2, \ldots, s \tag{2}$$

where $\rho = \lambda/\mu$, and $Q(t)$ is the number of customers in the system at time t. The probability of blocking BP is:

$$BP = \lim_{t \to \infty} P\{Q(t) = s\} = \frac{\rho^s/s!}{\sum_{i=0}^{s} \rho^i/i!} \tag{3}$$

and the average number of customers in the system (i.e. the average number of busy servers) is:

$$\lim_{t \to \infty} E[Q(t)] = E[Q] = (1 - BP)\rho. \tag{4}$$

The expression for $P_n(t)$ is independent of time, for sufficiently large t, due to the stationary arrival process.

In the case of the nonstationary arrival rate, $(d/dt)P_n(t)$ will converge to some value which is not necessarily zero at all time. In fact, $(d/dt)P_n(t)$ will converge to some function of time depending on the structure of $\lambda(t)$. Therefore, in order to obtain the queue-length distribution, one must solve the set of differential in (1). The solution to these differential equations is complex even for fairly small systems with special arrival rate functions $\lambda(t)$. For example, let us consider the simplest case where there is only one server in the system. Let $\lambda(t)$ and μ be the arrival rate function and service rate respectively. Then, the forward equations in (1) become:

$$\frac{d}{dt}P_0(t) = \mu P_1(t) - \lambda(t)P_0(t)$$

and

$$\frac{d}{dt}P_1(t) = \lambda(t)P_0(t) - \mu P_1(t)$$

where $P_0(0) = 1$ and $P_0(t) + P_1(t) = 1$, $t \geq 0$. This system can be reduced to solving a single differential equation:

$$\frac{d}{dt}P_1(t) + [\lambda(t) + \mu]P_1(t) = \lambda(t), \quad \text{with } P_1(0) = 0$$

which can be done as follows. Multiplying both sides by $e^{\int \lambda(\eta)+\mu \, d\eta}$ we have:

$$\frac{d}{dt}P_1(t)e^{\int \lambda(\eta)+\mu \, d\eta} + [\lambda(t)+\mu]e^{\int \lambda(\eta)+\mu \, d\eta}P_1(t) = \lambda(t)e^{\int \lambda(\eta)+\mu \, d\eta}$$

or

$$\frac{d}{dt}\left(P_1(t)e^{\int \lambda(\eta)+\mu \, d\eta}\right) = \lambda(t)e^{\int \lambda(\eta)+\mu \, d\eta},$$

$$P_1(t)e^{-\int \lambda(u)+\mu \, du} = \int \lambda(t)e^{\int \lambda(u)+\mu \, du} \, dt + K.$$

Given that $P_1(0) = 0$, we finally have the blocking probability, $BP(t)$:

$$BP(t) = P_1(t) = \int_0^t \lambda(u)e^{\int_0^u \lambda(\eta)+\mu \, d\eta} \, du \, e^{-\int \lambda(t)+\mu \, dt}.$$

In view of the difficulty in solving the nonstationary loss queue, various approximation methods have been developed. In this chapter, we review several of these approximation methods and we also present a new technique, the *fixed-point approximation* (FPA) method, which yields the mean number of customers and the blocking probability functions in a nonstationary loss queue. Numerical evidence points to the fact that the FPA method gives the exact solution.

The chapter is organised as follows. In Secs. 7.2 to 7.7 we describe the following approximation methods: the *simple stationary approximation* (SSA), the *stationary peakedness approximation* (PK), the *average stationary approximation* (ASA), the *closure approximation for nonstationary queues*, the *pointwise stationary approximation* (PSA), the *modified offered load approximation* (MOL). In Sec. 7.8, we present the fixed point approximation (FPA) method, and finally the conclusions are given in Sec. 7.9.

7.2. The Simple Stationary Approximation (SSA) Method

This method uses the average arrival rate of the nonstationary model to obtain the steady-state results. The average arrival rate for a cycle of length T is

$$\bar{\lambda} = \frac{1}{T} \int_0^T \lambda(t)\, dt. \tag{5}$$

Let $Q(t)$ be number of customers in the system at time t. Then, the steady-state distribution can be obtained using the expression:

$$P\{Q(t) = n\} = \frac{\rho^n/n!}{\sum_{i=0}^{s} \rho^i/i!}, \quad n = 0, 1, 2, \ldots, s, \text{ where } \rho = \frac{\bar{\lambda}}{\mu}$$

and the blocking probability, $BP(t)$, at time t is the Erlang loss formula with parameter (s, ρ), as follows:

$$BP(t) = P\{Q(t) = s\} = \frac{\rho^s/s!}{\sum_{i=0}^{s} \rho^i/i!}, \quad \text{for all } t.$$

The SSA method is simple and can be applied to a wide range of queuing systems. It provides a reasonable approximation for the nonstationary system with a weakly varying arrival rate. (An arrival rate is considered weakly varying over time if the arrival rate function remains within ±10% interval from the average arrival rate for all t.) However, this method noticeably underestimates the average performance measures of a nonstationary system with a strongly varying arrival rate. Green et al.[3] numerically investigated the level of nonstationarity at which this method provides a reasonable accuracy assuming a sinusoidal arrival rate function. The effect of nonstationarity with respect to amplitude, frequency of events and the size of the system (i.e. number of servers) was studied numerically. The authors showed that this method is applicable to relatively small systems (e.g. one or two servers) with small relative amplitude (e.g. less than 10%), and short cycle length (equivalently infrequent events).

Abdalla and Boucherie[4] used the SSA method to analyse a network of nonstationary loss queues. Consider a network of N independent loss queues, each with a time-dependent Poisson stream of external arrivals at rate $\lambda_i(t)$ and s_i servers, $i = 1, 2, \ldots, N$. Upon service completion at Queue i, a customer moves to node j with probability q_{ij} or it depart from the system with probability q_{i0}. Any external or internal arrival to Queue i finds all servers busy is lost.

The arrival rate to each queue in the network is averaged over time using (5). Then the probability that the system is in state $\mathbf{n}$ where $\mathbf{n} \in S$, $S = \{\mathbf{n} \in \mathbb{N}^N : 0 \leq n_i \leq s_i, i = 1, 2, \ldots, N\}$ is:

$$P(\mathbf{n}; t) = \prod_{i=1}^{N} \frac{\rho_i^{n_i}}{n_i!} \bigg/ \sum_{\mathbf{n} \in S} \prod_{j=1}^{N} \frac{\rho_j^{n_j}}{n_j!}.$$

The offered load of Queue j, ρ_j, satisfies the solution of the following traffic equations:

$$0 = \bar{\lambda}_j + \sum_{i=1}^{N} \mu_i q_{ij} \rho_i - \mu_j \rho_j, \quad j = 1, 2, \ldots, N.$$

It is worth mentioning that this method is an exact solution for loss networks with Markovian branching and stationary arrivals if and only the rates of Queue j, $j = 1, 2, \ldots, N$, in the network satisfy the following conditions:

$$\lambda_j = q_{jj} \rho_j \quad \text{and} \quad q_{ij} \rho_i = q_{ji} \rho_j, \quad j = 1, 2, \ldots, N.$$

The SSA underestimates the average performance measure of nonstationary systems even when the above two conditions are satisfied.

7.3. The Stationary Peakedness Approximation (PK) Method

The SSA method presented above does not consider the nonstationarity of the system. This can be done by using a non-Poisson stationary point process to approximate the time-dependent Poisson arrival process. Massey and Whitt[5] presented two approaches and used the heavy traffic peakedness to approximate the blocking probability of the nonstationary loss queue. (The peakedness is defined as the ratio of the variance to the mean of the steady-state number of customers in an infinite-server model with the same service time distribution and arrival process.)

To explain how this method works we consider a periodic Poisson arrival process with period T. The nonstationary arrival process is approximated by dividing the cycle T into n subintervals. It is assumed that the arrival rate at each subinterval is approximately constant. The arrival rate in any one subinterval is:

$$\lambda_k = \int_{(k-1)T/n}^{kT/n} \lambda(u)du, \quad 1 \leq k < n.$$

Then, the mean number of arrivals is:

$$\bar{\lambda}_n = \frac{1}{n}\sum_{k=1}^{n}\lambda_k = \frac{\bar{\lambda}T}{n}, \quad \text{where } \bar{\lambda} = \frac{1}{T}\int_0^T \lambda(u)du,$$

and its variance is:

$$\sigma_n^2 = \bar{\lambda}_n + \frac{1}{n}\sum_{k=1}^{n}(\lambda_k - \bar{\lambda}_n)^2.$$

Based on the above analysis, the overall arrival process M(t) in the interval $(0, T]$ is approximated by the stationary point process $\{N(t): t \geq 0\}$ with mean and variance:

$$n\bar{\lambda}_n = \bar{\lambda}T \quad \text{and} \quad n\sigma_n^2 = \bar{\lambda}T + \sum_{k=1}^{n}(\lambda_k - \bar{\lambda}_n)^2.$$

One may notice that the variance depends heavily on n. For example, for $n = 1$, $n\sigma_n^2 = \bar{\lambda}T$; while $n\sigma_n^2 \to \bar{\lambda}T$ as $n \to \infty$. Therefore, n should be an intermediate point to capture the variability in arrival process. Next, the peakedness c^2 for number of customers in the infinite server system, $Q(t)$, is calculated:

$$c^2 = \frac{Var[N(T)]}{E[N(T)]} = 1 + \frac{1}{\bar{\lambda}T}\sum_{k=0}^{n}(\lambda_k - \bar{\lambda}_n)^2.$$

Assuming that the arrival rate over each subinterval is constant, c^2 could be approximated as follows:

$$c^2 \approx 1 + \frac{1}{\bar{\lambda}T}\left(\frac{T}{n}\right)\int_0^T (\lambda(u) - \bar{\lambda})^2 \, du \approx 1 + \frac{1}{n\bar{\lambda}}\int_0^T (\lambda(u) - \bar{\lambda})^2 \, du.$$

It is always possible to rescale the problem to make the unit time equal to the mean service time. This means that $\mu = 1$. In this case, a good choice for n is to be equal to T. Thus,

$$c^2 \approx 1 + \frac{1}{\bar{\lambda}T} \int_0^T (\lambda(u) - \bar{\lambda})^2 \, du.$$

Then, c^2 is used to compute the heavy traffic peakedness of the nonstationary process. The heavy-traffic peakedness for an infinite-server system with exponential service distribution ($\mu = 1$) is:

$$z = 1 + \frac{c^2 - 1}{2} = 1 + \frac{1}{2\bar{\lambda}T} \int_0^T (\lambda(u) - \bar{\lambda})^2 \, du.$$

Finally, the approximate blocking probability for the M(t)/M/s/0 queue with time unit equals to the mean service time (i.e. $\mu = 1$) is given by the Erlang loss formula with updated number of servers s/z (s/z is integer) and updated offered load $\bar{\lambda}/z$ as follows:

$$BP(t) = P\{Q(t) = s\} = \frac{(\bar{\lambda}/z)^{s/z}/(s/z)!}{\sum_{i=0}^{s/z}(\bar{\lambda}/z)^i/i!}.$$

The average number of customers is:

$$E[Q(t)] = (1 - BP(t))\frac{\bar{\lambda}}{z}.$$

This method is a stationary approximation of the original system. In other words, the PK method finds non-Poisson stationary parameters that better approximate the time-dependent arrival process. Therefore, the resulting approximation with the new parameters is a time reversible process. Although this approximation does not provide a solution for the system as a function time, it provides a better approximation than the SSA method for the average measure of performance of nonstationary Erlang loss models.

7.4. The Average Stationary Approximation (ASA) Method

This method was introduced by Whitt[6] for loss queues with periodic arrival rates. This approximation starts by dividing the arrival rate cycle T into sub-intervals each of length τ, where τ is proportional or equal to the mean service time. The arrival rate $\lambda(t)$ over subinterval $[t-\tau, t]$ is taken to be equal to the average arrival rate during $[t-\tau, t]$ as follows:

$$\bar{\lambda}_k(t) = \frac{1}{\alpha\mu^{-1}} \int_{t_k-\alpha\mu^{-1}}^{t_k} \lambda(u)du, \quad t \in [t_k - \alpha\mu^{-1}, t_k], \quad \alpha\mu^{-1} = \tau.$$

The stationary results are used as a function of t and $\bar{\lambda}(t)$ to approximate the performance measures. Namely, the blocking probability, $BP(t,k)$, during sub-interval k is approximated as follows:

$$BP(t,k) = \frac{\rho_k^s/s!}{\sum_{i=0}^{s} \rho_k^s/i!}, \quad \rho_k = \frac{\bar{\lambda}_k(t)}{\mu}, \quad t \in [t_k - \alpha\mu^{-1}, t_k]$$

and the average number of customers, $E[Q(t,k)]$, during sub-interval k:

$$E[Q(t,k)] = (1 - BP(t,k))\frac{\bar{\lambda}(t)}{\mu}, \quad \text{for } t \in [t_k - \alpha\mu^{-1}, t_k].$$

Obviously, the performance measures are going to be step functions due to the discretisation of the arrival process. This method is simple to apply and it produces an insight into the behaviour of the performance measures over time. In addition, this method provides an exact solution for the $M(t)/D/\infty$ queue when $\alpha = 1$. This method depends mainly on the choice of the subinterval length (τ) which is strongly related to the choice of α. If α is chosen to be small when it should not be, the approximation will pick up more variability from the arrival process than needed. In contrast, if α is chosen to be large then this method will approach the stationary approximation which kills the variability of the performance measures over time.

7.5. The Closure Approximation for Non-Stationary Queues

This method reduces the number of differential equations needed to be solved by considering the differential equations of the mean and the variance of the number of customers in the system. In many systems, the equations for the mean and the variance involve more variables than the number of equations. Consequently, additional equations are required in order to obtain a unique solution.

Consider an $M(t)/M/1$ queue with arrival rate $\lambda(t)$ and service rate μ. The probability $P_n(t)$ of having n in the system at time t is given by the following set of differential equations:

$$\frac{d}{dt}P_0(t) = -\lambda(t)P_0(t) + \mu P_1(t),$$

$$\frac{d}{dt}P_n(t) = -(\lambda(t) + \mu)P_n(t) + \lambda(t)P_{n-1}(t) + \mu P_{n+1}(t), \quad n > 0.$$

(6)

Multiplying (6) by n and summing over all n gives:

$$\frac{d}{dt}E[n] = \sum_{n=0}^{\infty} n \frac{d}{dt}P_n(t) = \lambda(t) - \mu(1 - P_0(t)).$$

(7)

Multiplying (6) by n^2 and summing over all n gives:

$$\frac{d}{dt}E[n^2] = \sum_{n=0}^{\infty} n^2 \frac{d}{dt}P_n(t) = \lambda(t) - \mu(1 - P_0(t)) + 2E[n](\lambda(t) - \mu).$$

Hence, the variance is as follows:

$$\frac{d}{dt}Var[n] = \frac{d}{dt}E[n^2] - \frac{d}{dt}E[n]^2 = \lambda(t) + \mu P_0(t)(2E[n] + 1).$$

(8)

Equations (7) and (8) provide a system of two differential equations in three unknowns ($Var[n]$, $E[n]$ and $P_0(t)$). To obtain a unique solution using these equations an additional equation of $Var[n]$, $E[n]$ and $P_0(t)$ is required to bound the solution.

Rothkopf and Oren[7] consider the negative binomial distribution to provide a closure function for the $M(t)/M(t)/s$ system. The negative binomial with probability of success q and parameters n and r has the form:

$$p_n(q, r) = \binom{r + n - 1}{n} q^r (1 - q)^n, \quad n = 0, 1, 2, \ldots,$$

with mean $r(1-q)q^{-1}$ and variance is $r(1-q)q^{-2}$.

The negative binomial reduces to the geometric distribution if its parameters (q and r) are chosen such that:

$$Var[n] = E[n](1 + E[n]).$$

(9)

The number of customers in an M/M/1 system has a geometric distribution. Therefore, the parameters q and r can be chosen as functions of the mean and variance of the system so that the resulting negative binomial satisfies property (9). The new negative binomial distribution will have the following parameters:

$$q(t) = \frac{E[n]}{Var[n]} \quad \text{and} \quad r(t) = \frac{E[n]^2}{Var[n] - E[n]},$$

where $Var[n]$ and $E[n]$ are functions of time. The closure function $P_0(t)$ of (6) is obtained by setting n equals to zero in the negative binomial with the parameters $q(t)$ and $r(t)$:

$$P_0(t) = p_0(q(t), r(t)) = q(t)^{r(t)}.$$

Similarly, the mean and variance of $M(t)/M/s$ system are:

$$\frac{d}{dt}E[n] = \lambda(t) - \mu s + \mu \sum_{n=0}^{s-1}(s-n)P_n(t) \qquad (10)$$

and

$$\frac{d}{dt}Var[n] = \lambda(t) + \mu s - \mu \sum_{n=0}^{s-1}(2E_t[n]+1-2n)(s-n)P_n(t). \qquad (11)$$

The closure functions $(P_n(t)$ for $n = 0, 1, 2, \ldots, s-1)$ of (10) and (11) are obtained by evaluating the negative binomial distribution described earlier at $n = 0, 1, 2, \ldots, s-1$.

This method provides an exact solution for the stationary $M/M/1$ queue and a very good approximation for the $M(t)/M/1$ queue due to the fact that the stationary system has a geometric steady-state solution. However, for the $M(t)/M/s$ queue the error of the approximation increases very quickly as the number of servers increases. Rothkopf and Oren[7] provided an error correction term to improve the accuracy of the approximation for the $M(t)/M/s/s$ queue.

7.6. The Pointwise Stationary Approximation (PSA) Method

This method is based on the idea that the nonstationary loss queue approximately behaves like a stationary model at each instance of time. Thus, the steady-state results of the stationary loss queue can be used to approximate the nonstationary loss queue at each point on time. This method was first introduced by Grassman[8] in 1983 as a way of constructing an upper bound on the expected number of customers in the queue. Green et al.[3] showed numerically that PSA gives an upper bound on the expected number of customers in the system and probability of delay, if the maximum traffic intensity is strictly less than one. In addition, Green and Kolesar[10] used the PSA method to approximate the steady-state average performance measures of the periodic $M(t)/M/s/s$ queue.

Consider a stationary loss queue with arrival rate λ and service rate μ. Let $Q(t)$ be number of customers in the system at time t. Then, the probability P_n that there are n customers in the system is:

$$P_n = \lim_{t\to\infty} P\{Q(t) = n\} = \frac{\rho^n/n!}{\sum_{i=0}^{s} \rho^i/i!},$$

$$\rho = \frac{\lambda}{\mu}, \quad n = 0, 1, 2, \ldots, s.$$

The probability of blocking BP is:

$$BP = \lim_{t\to\infty} P\{Q(t) = s\} = \frac{\rho^s/s!}{\sum_{i=0}^{s} \rho^i/i!},$$

and the average number in the system (i.e. average number of busy servers) is:

$$\lim_{t\to\infty} E[Q(t)] = E[Q] = (1 - BP)\rho.$$

In the PSA method, the time dependent-steady state distribution of the nonstationary Erlang loss system, given that the arrival rate is $\lambda(t)$ and service rate is μ, is calculated as follows:

$$P_n(t) = \frac{\rho(t)^n/n!}{\sum_{i=0}^{s} \rho(t)^i/i!}, \quad \rho(t) = \frac{\lambda(t)}{\mu} \quad \text{and} \quad n = 0, 1, 2, \ldots, s$$

the time-dependent steady-state blocking probability is

$$BP(t) = Ps(t) = \frac{\rho(t)^s/s!}{\sum_{i=0}^{s} \rho(t)^i/i!},$$

and the time-dependent steady-state average number in the system is:

$$E[Q(t)] = (1 - BP(t))\rho(t)$$

where $\rho(t) = \lambda(t)/\mu$.

The PSA method can be easily generalised to most of the queuing systems, as long as $\rho < 1$ for all t is required for the stability of the equivalent stationary system.

An important factor that affects the accuracy of the PSA is the arrival rate function. The PSA method will provide a good approximation as the arrival rate increases. For example, consider two nonstationary loss queues with sinusoidal arrival rate function $\lambda(t) = \bar{\lambda} + \beta\sin(\gamma T)$ where $\bar{\lambda}$ is the average arrival rate, β is the amplitude and γ is the frequency set equal to $2\pi/T$, T being the cycle length. Let $\{\lambda(t) = 5 + 2.5\sin(t), \mu = 0.5, s = 10\}$ and $\{\lambda(t) = 20 + 10\sin(t), \mu = 2, s = 10\}$ be the parameters of loss queues 1 and 2 respectively. Then according to PSA, the time-dependent offered load for both systems is

$$\rho(t) = \frac{\lambda(t)}{\mu} = 10 + 5\sin(t).$$

Although both systems have the same offered load, PSA will provide a better approximation for loss Queue 2, since it needs shorter time to reach the steady state due to the higher arrival and service rates. This means that loss Queue 2 will behave more like a stationary system within reasonably small interval of time than Queue 1.

In Fig. 7.1, we plot the exact time-dependent average number $E[Q(t)]$ for loss queue 1, 2 (labelled "Queue 1" and "Queue 2" respectively in Fig. 7.1) and the PSA values as a function of time t. As expected, PSA provides better approximation for loss Queue 2 than for loss Queue 1.

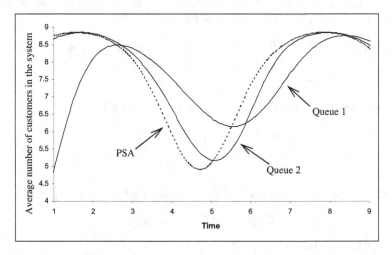

Fig. 7.1. PSA and exact values of the average number of customers in loss queues 1 and 2.

As can be seen in Fig. 7.1, the PSA method overestimates the peak of the average number of customers. In addition, the PSA peak lags the peak of the average number of customers. (The same also applies to other performance measures.) These two problems become negligible as the arrival and service

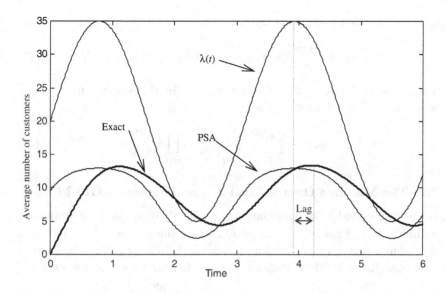

Fig. 7.2. Exact and PSA values of the average number of customers.

rates increase. Whitt[6] showed that the PSA solution for an $M(t)/M(t)/s$ queue is asymptotically correct as the rates increase.

Green and Kolesar[12,13] proposed the *Simple Peak Hour Approximation* (SPHA) technique for the computation of average performance measures of periodic systems during the peak period. SPHA starts by obtaining the measure of interest, say $X(t)$, using the PSA method. Next, the peak time t^* at which the $X(t)$ achieves its maximum is determined. The average of $X(t)$ over the interval $[a, b]$ where t^* is the center of the interval is the SPHA value for $X(t)$.

SSA and PSA can be seen as two extreme cases of averaging out the arrival rate. The SSA method averages the arrival rate $\lambda(t)$ over a period of time equal to the cycle length T, whereas the PSA method uses the average arrival rate over an infinitesimally small interval.

Finally, the PSA method has also been used to analyse a network nonstationary loss queues, see Massey and Whitt[14] and Abdalla and Boucherie[4]. The same system of traffic equations is used as in the SSA method only the arrival rates are taken to be functions of time instead of averages. That is, the time-dependent offered loads $\rho_j(t)$ are obtained by solving the following system of traffic equations in time t:

$$0 = \lambda_j(t) + \sum_{i=1}^{N} \mu_i q_{ij} \rho_i(t) - \mu_j \rho_j(t); \quad j = 1, 2, \ldots, N. \qquad (12)$$

Then, the probability, $P(\mathbf{n}; t)$, of having $\mathbf{n}$ in the system, for $\mathbf{n} \in S$, $S = \{\mathbf{n} \in \mathbb{N}^N : 0 \le n_i \le s_i; i = 1, 2, \ldots, N\}$, is

$$P(\mathbf{n}; t) = \prod_{i=1}^{N} \frac{\rho_i(t)^{n_i}}{n_i!} \Big/ \sum_{\mathbf{n} \in S} \prod_{j=1}^{N} \frac{\rho_j(t)^{n_j}}{n_j!}. \qquad (13)$$

7.7. The Modified Offered Load Approximation (MOL) Method

Let us first consider the stationary loss queue M/M/s/s and the stationary infinite server queue M/M/∞. Let $Q_\infty(t)$ be the number of customers in the infinite server queue at time t. The probability P_n that there are n customers in an M/M/∞ with an arrival rate λ and a service rate μ is:

$$P_n = \lim_{t \to \infty} P\{Q_\infty(t) = n\} = \frac{\rho^n}{n!} e^{-\rho},$$

where $\rho = \lambda/\mu$. Likewise, let $Q(t)$ be number of customers in the loss queue with s servers at time t. The probability P_n that there are n customers in an M/M/s/0 with an arrival rate λ and a service rate μ, is:

$$P_n = \lim_{t \to \infty} P\{Q(t) = n\} = \frac{\rho^n/n!}{\sum_{i=0}^{s} \rho^i/i!},$$

where $\rho = \lambda/\mu$ and $n = 0, 1, 2, \ldots, s$.

Another way to obtain the stationary distribution of the M/M/s/0 queue is to use the fact that the M/M/s/0 queue is a truncated process of an M/M/∞ queue which is a reversible Markov process. Since the arrival rates are time invariant we dropped the time variable from the random variables. Then, we have:

$$P\{Q = n\} = P\{Q_\infty = n | Q_\infty < s\} = \frac{e^{-\rho}(\rho^n/n!)}{e^{-\rho} \sum_{i=0}^{s} \rho^i/i!} = \frac{\rho^n/n!}{\sum_{i=0}^{s} \rho^i/i!}.$$

In the M(t)/M/∞ queue, the rate of change in the average number of customers at time t is equal to the difference between the arrival rate and the departure rate due to the Markovian property. That is,

$$\frac{d}{dt} E[Q_\infty(t)] = \lambda(t) - \mu E[Q_\infty(t)].$$

Recall that there is always an idle server for each arriving customer to the $M(t)/M/\infty$ queue. This means that no customers are lost and all customers in the system at time t are being served. Therefore, the average number of customers at time t is equal to the average number of customers in the system at time t which equals to the offered load $\rho(t)$. We have the following differential equation for $\rho(t)$:

$$\frac{d}{dt}\rho(t) = \lambda(t) - \mu\rho(t). \tag{14}$$

Analogous to the stationary queues, one can approximate the $M(t)/M/s/0$ by truncating the $M(t)/M/\infty$ queue. This method is called the *modified offered load method* (MOL). The MOL approximation was first developed by Jagerman[15] in 1975. The probability $P_n(t)$ that there are n customers in the system using MOL is:

$$P_n(t) \approx P\{Q_\infty(t) = n | Q_\infty(t) < s\} = \frac{\rho(t)^n/n!}{\sum_{i=0}^{s}\rho(t)^i/i!}, \quad n = 0, 1, 2, \ldots, s$$

where $\rho(t)$ is determined from (14).

The truncated $M/M/\infty$ queue provides an exact solution to the $M/M/s/0$ queue due to the reversibility property. In the case of nonstationary arrival process, the reversibility property is lost and hence the truncated $M(t)/M/\infty$ will not provide an exact solution to the $M(t)/M/s/0$. Massy and Whitt[16] developed analytical bounds on the error between the MOL approximation and the exact solution of the $M(t)/M/s/0$ system.

The MOL method can be seen as averaging out the arrival rate over an interval that depends on the mean and the distribution of the service time. This is in contrast to the SSA method where the arrival rate $\lambda(t)$ is averaged over the cycle length T, and the PSA method where the arrival rate is averaged over an infinitesimally small interval.

The $M(t)/M/s/s$ behaves like $M(t)/M/\infty$ as the blocking probability gets smaller. In view of this, the MOL method provides a good approximation for the $M(t)/M/s/s$ as long as the system has a small blocking probability. Experiments showed that the actual blocking probability of the $M(t)/M/s/s$ queue should be less than 0.1 in order for the MOL to provide a good approximation. As expected, the MOL underestimates the blocking probability of a loss queue with a high load, i.e. when the exact blocking probability is high.

The MOL method provides a good estimation for the peak time for a loss queue with a small blocking probability. This is due to the fact that the MOL method is sensitive to the service process through its mean and distribution. The PSA method depends on the service distribution only through its mean. As a result, PSA appears to lag the actual performance measures values of the system. Figure 7.3 shows the exact, PSA, and MOL values of the average number of customers in an $M(t)/M/s/s$ queue with $\lambda(t) = 20 + 15\sin(2t)$, $\mu = 2$ and $s = 15$.

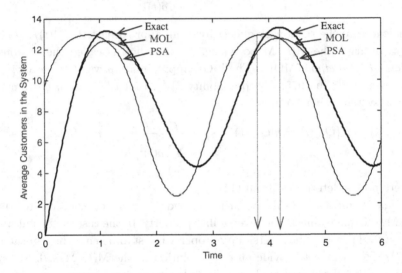

Fig. 7.3. Exact, PSA and MOL values of the average number of customers.

The MOL method has also been used to analyse networks of nonstationary loss queues. Whitt[17] described how the MOL method can be used in a decomposition algorithm for the analysis of a nonstationary loss network. Jennings and Massey[18] used the MOL method to analyse time-dependent circuit-switched networks. Abdalla and Boucherie[4] applied the MOL method to a mobile communication network with time-varying arrival rates and redialing. The authors established an exact expression for the error in the MOL approximation as well as bounds on the error.

The MOL method is used to analyse a network of nonstationary loss queues as follows. Consider a network consisting of N independent loss queues, each with a time dependent Poisson stream of external arrivals at rate $\lambda_i(t)$ and s_i servers, $i = 1, 2, \ldots, N$. Upon service completion at Queue i, a customer moves to Queue j with probability q_{ij} or it depart from the system with probability q_{i0}. Any external or internal arrival to

Queue i finds all servers busy is lost. The traffic equations of the equivalent network consisting of nonstationary infinite server queues are first solved in order to obtain the time-dependent offered loads $\rho_i(t)$, $i = 1, 2, \ldots, N$. This system of traffic equations is the same as in (12) used in the PSA method except that the $(d/dt)\rho_i(t)$, $i = 1, 2, \ldots, N$, are not taken to be zero. That is, the time dependent offered loads $\rho_j(t)$ are obtained by solving the following system of differential equations in time t:

$$\frac{d}{dt}\rho_j(t) = \lambda_j(t) + \sum_{i=1}^{N} \mu_i q_{ij} \rho_i(t) - \mu_j \rho_j(t); \quad j = 1, 2, \ldots, N.$$

Then, for $\mathbf{n} \in \{\mathbf{n} \in \mathbb{N}^N : 0 \leq n_i \leq s_i; i = 1, 2, \ldots, N\}$

$$P(\mathbf{n}; t) = \prod_{i=1}^{N} \frac{\rho_i(t)^{n_i}}{n_i!} \bigg/ \sum_{n \in S} \prod_{j=1}^{N} \frac{\rho_j(t)^{n_j}}{n_j!}.$$

It is worth mentioning that the $M_t/M/s/0$ behaves like $M_t/M/\infty$ as the blocking probability gets smaller. Then, it is expected that the MOL method will provide a good approximation for the nodes in the system that has a small blocking probability.

7.8. The Fixed Point Approximation (FPA) Method

The *fixed point approximation* (FPA) method was proposed by Alnowibet and Perros[19]. This method calculates numerically the time-dependent mean number of customers and blocking probability functions in a nonstationary multi-rate loss queue. Experimental results showed that the FPA algorithm provides an exact solution. The FPA method has also been extended to nonstationary queuing networks of multi-rate loss queues, see Alnowibet and Perros[19], and nonstationary queuing networks with population constraints, see Alnowibet and Perros[20]. In this chapter, we describe the FPA method for the analysis of the nonstationary (single-class) loss queue. Consider a loss queue $M(t)/M/s/s$ with a Poisson arrival process with timedependent rate $\lambda(t)$. The time-dependent average number of customers $E[Q(t)]$ in an $M(t)/M/s/s$ queue can be expressed as the difference between the effective arrival rate and the departure rate at time t. We have:

$$\frac{d}{dt}E[Q(t)] = \lambda(t)(1 - BP(t)) - [\mu P\{Q(t) = 1\} + 2\mu P\{Q(t) = 2\} + \cdots \\ + s\mu P\{Q(t) = s\}]$$

where $BP(t)$ is the blocking probability at time t. The above equation can be written as follows:

$$\frac{d}{dt}E[Q(t)] = \lambda(t)(1 - BP(t)) - \mu E[Q(t)]. \tag{15}$$

We note that the time-dependent mean number of customers is given by the expression: $E[Q(t)] = \rho(t)(1 - BP(t))$, from which we have the following expression for the offered load $\rho(t)$:

$$\rho(t) = \frac{E[Q(t)]}{1 - BP(t)}, \tag{16}$$

$$BP(t) = \frac{\rho(t)^s/s!}{\sum_{i=0}^{s}\rho(t)^i/i!}. \tag{17}$$

Using expressions in (15), (16), and (17), we can calculate the blocking probability iteratively. The steps of the algorithm are as follows:

1. Choose an appropriate Δt, final time T_f and tolerance ε.
2. Choose initial conditions for $E[Q(t)]$. Set $E[Q(0)] = 0$.
3. Evaluate $\lambda(t)$ at $t = 0, \Delta t, 2\Delta t, \ldots, T_f$.
4. Start with an initial blocking probability $BP^0(t) = 0$, $t = 0, \Delta t, 2\Delta t, \ldots, T_f$.
5. Set the iteration counter $k = 0$.
6. Solve numerically for $E[Q^k(t)]$ using the following equation:

$$E[Q^k(t + \Delta t)] = E[Q^k(t)] + \lambda(t)(1 - BP^k(t))\Delta t - \mu E[Q^k(t)]\Delta t.$$

7. Calculate $\rho^k(t) = \dfrac{E[Q^k(t)]}{1 - BP^k(t)}$, $t = 0, \Delta t, 2\Delta t, \ldots, T_f$.

8. Update blocking probability $BP^{k+1}(t) = \dfrac{[\rho^k(t)]^s/s!}{\sum_{i=0}^{s}[\rho^k(t)]^i/i!}$,

$t = 0, \Delta t, \ldots, T_f$.

9. If $\|BP^k(t) - BP^{k+1}(t)\| < \varepsilon$, then $BP^k(t)$ has converged and the algorithm stops. Else, set $k = k + 1$ and go to step 6.

The algorithm does not require a closed-form expression for the arrival rate function. It only needs that the arrival rate function be defined at time points equally spaced by Δt. In view of this, any periodic arrival rate function can be used irrespective of whether we know its closed-form or not.

Since this algorithm discretises the arrival rate function, the continuity and differentiability properties of the arrival rate function are not necessary.

We also note that the algorithm can be easily extended to the case where the service rate is also time-dependent by simply defining the service rate as a vector corresponding to the same time points used for the arrival rate function.

In all the experiments the FPA results were very close to the exact numerical results or within the simulation confidence intervals. This lead us to the conjecture that (17) for the nonstationary blocking probability used in the FPA method is in fact correct. However, due to the discretisation process, the FPA and the exact numerical results never matched, which prevented from establishing beyond doubt the correctness of (17). As an example, let us consider a loss queue with a sinusoidal arrival rate function $\lambda(t) = \bar{\lambda} + \beta \sin(\gamma t)$, where $\bar{\lambda} = 20$, $\beta = 15$ and $\gamma = 2$, $s = 10$, and the service rate $\mu = 1$. The FPA method was applied with tolerance $\varepsilon = 0.01$ for different values of Δt. The average absolute error between the exact and the FPA solutions for the blocking probability and the average number of customers are given in Table 7.1. As can be seen the absolute error decreases as Δt decreases. Due to the CPU and memory limitations, it was not possible to consider Δt values less than 0.0001.

Table 7.1. Average absolute error of FPA as $\Delta t \to 0$.

	$\Delta t = 0.1$	$\Delta t = 0.05$	$\Delta t = 0.01$	$\Delta t = 0.005$	$\Delta t = 0.001$	$\Delta t = 0.0005$	$\Delta t = 0.0001$
$\Delta BP(t)$	0.269	0.0165	0.0123	0.0118	0.0115	0.0114	0.0114
$E[Q(t)]$	0.0992	0.692	0.0507	0.0474	0.0447	0.0445	0.0445

7.9. Conclusions

The loss queue has been extensively studied in the stationary case, i.e., assuming that the arrival rate and the service rate are time invariant. The nonstationary loss queue, where the arrival rate is time-dependent is also of interest, since the arrival rate in most communication systems varies over time. In view of the difficulty in solving the nonstationary loss queue, various approximation methods have been developed. In this chapter, we reviewed the following approximation methods: the simple stationary approximation (SSA), the stationary peakedness approximation (PK), the average stationary approximation (ASA), the closure approximation for nonstationary queues, the pointwise stationary approximation (PSA), and the modified offered load approximation (MOL). We also presented a new technique, referred to as the *fixed-point approximation* (FPA) method, which yields the mean number of customers and the blocking probability functions in a nonstationary loss queues. Numerical evidence points to the fact that the FPA method gives the exact solution.

References

1. Washington, A. N. and Perros, H. G., "Call blocking probabilities in a traffic groomed tandem optical network", Special issue dedicated to the memory of Professor Olga Casals, Blondia and Stavrakakis (Eds.) *Journal of Computer Networks*, **45** (2004), 281-294.
2. Battestilli, L. and Perros, H. G., "End-to-End Burst Probabilities in an OBS Network with Simultaneous Link Possession" Workshop on OBS, BroadNets 2004.
http://www.csc.ncsu.edu/faculty/perros//recentpapers.html.
3. Green, L. V., Kolesar, P. J. and Svoronos, A., "Some Effects of Nonstationarity On Multiserver Markovian Queues Systems", *Operations Research*, **39** (1991), 502-511.
4. Abdalla, N. and Boucherie R. J., "Blocking Probabilities in Mobile Communications Networks with Time-Varying Rates and Redialing Subscribers", *Annals of Operations Research*, **112** (2002), 15-34.
5. Massey, W. A. and Whitt W., "Stationary-Process Approximation for the Nonstationary Erlang Loss Model", *Operations Research*, **44** (1996), 976-983.
6. Whitt W., "The Pointwise Stationary Approximation for Mt/Mt/s Queues is Asymptotically Correct as the Rates Increase", *Management Sciences*, **37** (1991), 307-314.
7. Rothkopf, M. H. and Oren, S. S., "A Closure Approximation for the Nonstationary M/M/s Queue", *Management Sciences*, **25** (1979), 522-534.
8. Grassmann, W., "The Convexity of the Mean Queue Size of the M/M/c Queue with Respect to the Traffic Intensity", *Journal of Applied Probability*, **20** (1983), 916-919.
9. Green, L.V. and Kolesar, P. J., "The Pointwise Stationary Approximation for Queues with Nonstationary Arrivals", *Management Sciences*, **37** (1991), 84-97.
10. Green, L.V. and Kolesar, P. J., "The Lagged PSA for Estimating Peak Congestion in Multiserver Markovian Queues with Periodic Arrival Rates", *Management Sciences*, **43** (1997), 80-87.
11. Green, L.V. and Kolesar, P. J., "On the Accuracy of the Simple Peak Hour Approximation for Markovian Queues", *Management Sciences*, **41** (1995), 1353-1370.
12. Massey, W. A. and Whitt W., "Networks of Infinite-Server Queues with Nonstationary Poisson Input", *Queuing Systems*, **13** (1993), 183-251.
13. Jagerman, D. L., "Nonstationary Blocking in Telephone Traffic", *The Bell*

System Technical Journal, **54** (1975), 626-661.
14. Massey, W. A. and Whitt W., "An Analysis of the Modified Offered-load Approximation for the Nonstationary Loss Model", *Annals of Applied Probability*, **4** (1994), 1145-1160.
15. Whitt, W., "Decomposition Approximation for Time-Dependent Markovian Queuing Networks", *Operations Research Letters*, **24** (1999), 97-103.
16. Jennings, O. B. and Massey, W. A., "A Modified Offered Load Approximation for Nonstationary Circuit Switched Networks", *Telecommunication Systems*, **7** (1997), 229-251.
17. Alnowibet, K. and Perros, H.G., "Nonstationary Loss Queues and Loss Queueing Networks",
http://www.csc.ncsu.edu/faculty/perros//recentpapers.html
18. Alnowibet, K. and Perros, H., "Nonstationary Analysis of Circuit-Switched Communication networks"
http://www.csc.ncsu.edu/faculty/perros

CHAPTER 8

Stabilisation Techniques for Load-Dependent Queuing Networks Algorithms

Giuliano Casale and Giuseppe Serazzi

Politecnico di Milano, Dip. di Elettronica e Informazione
Via Ponzio 34/5, I-20133 Milano, Italy
{giuliano.casale, giuseppe.serazzi}@polimi.it

Product-Form (PF) queuing networks are one of the most popular modelling techniques for evaluating the performances of computing and telecommunication infrastructures. Several computational algorithms have been developed for their exact solution. Unlike the algorithms for load-independent models, the ones for models with queue-dependent servers, either with single class or multiclass workloads, are numerically unstable. Furthermore, existing numerical stabilisation technique for multiclass load-dependent models are not always efficient. The search for such result is motivated by the complexity of nowadays systems, which often require the use of load-dependent models.

In this work we review numerical instabilities of computational algorithms for product-form queuing networks. Then, we propose a general solution to the problem based on arbitrary precision arithmetics. Finally, we discuss a new specialised stabilisation technique for two class load-dependent models.

8.1. Introduction

Closed queuing network models[10] have played an important role in the performance evaluation of complex computer and telecommunication systems[17]. Their applications[14,4,13] span over several types of performance evaluation studies, such as system tuning, optimisation and capacity planning.

Among the most important reasons of this success is the availability of a simple product-form expression of the exact steady-state probability distribution of network states[3]. The queuing networks that have such closed form expression are referred to as Product-Form Queuing Networks (PFQN). This property has led to the definition of efficient exact algorithms, including the Convolution Algorithm[6], the Mean Value Analysis[20] (MVA) and algorithms that work by recursion on the number of customer classes, e.g. RECAL[8]. An important feature of the MVA algorithm, compared to the others, is that it grants the numerical stability of the computation, meaning that: (i) the quantities computed at each step always lie within the floating-point range allowed by the computer architecture; (ii) no significant precision loss due to numerical round-offs is suffered. Unfortunately, this interesting property holds only for *load-independent* models, i.e. models where the rate at which the jobs are served are processed at each server is assumed to be constant. Indeed, exact solution algorithms for load-dependent models, i.e., models comprising servers with queue-dependent service rates, either with a single class or multiclass customers, frequently exhibit numerical instabilities.

Nevertheless, models of Web architectures contain a large number of heterogenous components that often require the use of load-dependent servers. For example, approximations of subsystems including population constraints (e.g. maximum number of HTTP connections) may be obtained with load-dependent PF flow-equivalent servers[14]. In this work we review numerical instabilities of computational algorithms for PFQN. Then, we propose a general solution to the problem based on arbitrary precision arithmetics. Finally, we consider a stabilisation technique for two class load-dependent models. The paper is organised as follows: Section 8.2 gives preliminary concepts on numerical instabilities and queuing networks. Section 8.3 explores instabilities in the Convolution Algorithm and in the load-dependent MVA. Section 8.4 discusses two new stabilisation techniques. Finally, Sec. 8.5 concludes the chapter.

8.2. Preliminaries

8.2.1. *Numerical Exceptions*

In this section we give a brief overview of numerical instabilities in floating-point computations. In actual computer architectures a floating point number[1] is represented using three elements: a sign, an exponent e and a mantissa m. For instance, the number $x = -1.04 \cdot 10^{-12}$ has negative sign, mantissa $m = 1.04$ and exponent $e = -12$. Looking at the binary representation, the sign takes a single bit while the remaining bits, whose number depends on the data type, are split between the mantissa and the exponent. For instance, the double precision arithmetics is implemented with 53 bits (i.e. 16 decimal digits) in the mantissa and 11 bits in the exponent, with the representable numbers ranging approximately in $[10^{-308}, 10^{+308}]$. Despite this floating-point range may appear sufficient for most algorithms, numerical exceptions frequently arise. In general, an algorithm $A(\mathbf{x})$ for computing a function $f(\mathbf{x})$ is said to be *numerically stable* if

$$A(\mathbf{x}) \approx f(\mathbf{x} + \epsilon),$$

where ϵ is a small perturbation of the input vector. Informally, this definition wishes to capture that arbitrary small perturbations of the input parameters, like those implied by limited machine accuracy, should not affect macroscopically the output of $A(\mathbf{x})$. Unfortunately, several numerical exceptions may prevent $A(\mathbf{x})$ from returning a correct result:

- *round-off errors*: errors associated with the existence of a limited number of digits for representing the mantissa. For instance, some irrational numbers, like $\pi = 3.1415\ldots$, cannot be represented with a finite number of digits. Thus, rounding schemes (like toward zero or nearest-value rounding[1]) are required to achieve the best possible approximation. As an example of the problems that may arise with double precision arithmetics, let $y = 10^{+23}$ and $x = 1$, define $w = (y-y)+x$ and $z = (y+x)-y$, then due to round-off errors we get $w = 1$ and $z = 0$! Note that even if rounding schemes may result in small errors for a single operation, e.g. of the order of 10^{-16} for double precision arithmetics, they can accumulate during execution and yield macroscopic errors on $A(\mathbf{x})$;
- *underflow/overflow errors*: underflow and overflow errors occur when the number to be represented is outside the allowed floating-point range. The limiting factor in this case is the number of digits of the exponent. For instance, let $x = 10^{+257}$ and $y = 10^{+52}$, it is straightforward to see that the product $z = xy = 10^{+309}$ exceeds the IEEE double precision range. Thus it is internally represented as an infinite (`Inf`) that cannot be used in other arithmetics operations.

The effect of the above exceptions on computational algorithms for product-form queuing networks are discussed in the following sections.

8.2.2. Closed Product-Form Queuing Networks

From now on, we consider closed PFQN composed of M servers that process the request of N customers. We assume that $M_Q \geq 0$ queue-dependent servers are present in the network. Customers are partitioned in R classes, each with a constant population N_r. Network population is therefore described by a population vector $\mathbf{N} = (N_1, \ldots, N_R)$ and we define $N = \sum_{r=1}^{R} N_r$. The vector $\mathbf{0} = (0, \ldots, 0)$ denotes an empty population; $\mathbf{1}_r$ is, instead, the r-dimensional unit vector. We denote L_{ir} the class r loading at server i, i.e., the average time spent by class r customers at server i if no queuing is experienced. Note that this term also accounts for network topology[9]. For queue-dependent servers, the rate at which request are processed when n customers queue at server i is expressed through the capacity function[5] $c_i(\mathbf{n})$. Later, we refer to the following mean performance indices:

$$X_r(\mathbf{N}) = \text{mean throughput of class } r$$
$$Q_{ir}(\mathbf{N}) = \text{mean queue length of class } r \text{ at server } i$$

Finally, if not otherwise specified, $i, m = 1, .., M$ and $r, s = 1, .., R$ should be intended as servers and customer classes indices, respectively.

8.3. Numerical Instabilities in PFQN Algorithms

8.3.1. *Convolution Algorithm*

The Convolution Algorithm[6,18] allows to recursively compute the normalisation constant $G(\mathbf{N})$ for load-dependent and load-independent multiclass PFQN. Simple analytical expression allow to derive any $X_r(\mathbf{N})$ or $Q_{ir}(\mathbf{N})$ from the knowledge of a set of normalisation constants. For instance, the throughput can be computed using the following formula:

$$X_r(\mathbf{N}) = \frac{G(\mathbf{N} - \mathbf{1}_r)}{G(\mathbf{N})} \quad (1)$$

The Convolution Algorithm for load-dependent models is based on the following recursion:

$$G(\mathbf{N}) = G(M, \mathbf{N}) = \sum_{\mathbf{n} \leq \mathbf{N}} F_M(\mathbf{n}) G(M-1, \mathbf{N} - \mathbf{n}) \quad (2)$$

with initial condition $G(1, \mathbf{N}) = F_1(\mathbf{N})$ and where $\mathbf{n} = (n_1, \ldots, n_R)$ is a vector of positive integers. In the above formula, the term $G(M-1, \mathbf{N} - \mathbf{n})$ refers to the normalisation constant of the *M-complement system*, i.e. the network obtained from the original one by removing the server indexed with M. Instead, the *product-form factor* (*PF factor*) $F_M(\mathbf{n})$ for queue M accounts for the probability that server M contains a population $\mathbf{n}^3$. Alternative forms of load-dependence are possible[14,5]. Let $|\mathbf{N}| \triangleq \prod_r (N_r + 1)$ be the number of populations vectors $\mathbf{n}$ such that $\mathbf{0} \leq \mathbf{n} \leq \mathbf{N}$. Let $M \geq 2$, then the time complexity of a recursive computation of (2) for a network consisting of queue-dependent stations ($M = M_Q$) is bounded by $O(M_Q |\mathbf{N}|^2)$, while the space complexity is $O(2 |\mathbf{N}|)$. Multiclass models that contain both open and closed classes can be solved with a slightly modified version of (2) and with a similar computational effort. Note that, for load-independent models, (2) can be simplified to

$$G(\mathbf{N}) = G(M, \mathbf{N}) = G(M - 1, \mathbf{N}) + \sum_r L_{mr} G(M, \mathbf{N} - \mathbf{1}_r) \quad (3)$$

and its solution requires $O(M \prod_r (N_r+1))$ time and $O(2 \prod_r (N_r+1))$ space. As discussed in several works[15], a major problem connected to the use of normalisation constants is that their growth can easily led to buffer overflows or underflows. For example, for $M = R = 3$, $N_1 = N_2 = N_3 = 100$ and all loadings L_{ir} equal to 100, the normalisation constant has value $G = 1.7 \cdot 10^{445}$, which exceeds the IEEE double precision range. Scaling techniques, described in the next section, are effective in addressing the problem.

8.3.1.1. Static and Dynamic Scaling Techniques

Several solutions have been proposed in literature for range extension problems on the basis of the following observation: assume that, for each class r, the loadings L_{ir} of all servers are initially scaled by some common factor β_r, then the normalisation constant can be rescaled through the following relation:

$$G(\alpha, \mathbf{N}) = r(\beta, \mathbf{N}) G(\beta_1, .., \beta_R, \mathbf{N}) \quad (4)$$

where $\alpha = (\alpha_1, .., \alpha_R)$ and $\beta = (\beta_1, .., \beta_R)$ are two *scaling vectors* and

$$r(\beta, \mathbf{N}) = \prod_r^R \left(\frac{\alpha_r}{\beta_r}\right)$$

is a conversion function between normalisation constants employing different scaling vectors. For instance, for the three servers and three classes example considered before, assume that the class-1 loadings L_{k1} are all multiplied by $\alpha_1 = 0.001$. Let $\alpha = (0.001, 1, 1)$ and $\beta = (1, 1, 1)$, then the resulting normalisation constant $G(\alpha, \mathbf{N})$ is computed as

$$G(\alpha, \mathbf{N}) = 10^{-300} G(\beta, \mathbf{N}) = 1.7 \cdot 10^{145}$$

that now lies in the range $[10^{-308}, 10^{+308}]$. Mean performance measures are easily obtained from the computed normalisation constants and the associated scaling vectors. Thus, a non-negligible memory overhead may be associated with the use of scaling techniques. However, this may be limited by employing for all classes a same scaling factor $\alpha = \alpha_1 = \ldots = \alpha_R$ that can be computed as a function of N and of the floating-point range[15].

A problem that may arise in the implementation of scaling techniques is the choice of an initial $\alpha = \alpha_1 = \ldots = \alpha_R$ everytime the normalisation constant of a new population is evaluated. Empirically, if in (3) the scaling factor α is initially assumed to be $\alpha = 1$, then the resulting computation remains numerically unstable because $r(\alpha, \mathbf{N})$ may produce buffer overflows/underflows for large populations. Thus, α should be evaluated as a function of the scaling factors of the normalisation constants involved in (3). We observed that if α is set to the maximum of the scaling factors β, then no stability problems arise.

Following a simpler approach, it also possible to scale the loadings L_{ir} before algorithm execution. This form of scaling is commonly referred to as *static scaling*, while the one described above is known as *dynamic scaling*. Previous works[19,15] have shown the effectiveness of static scalings for load-independent models. In particular, the following scaling can often limit the growth of $G(\mathbf{N})$ in multiclass models:

$$L_{ir} \leftarrow \left(\frac{\sum_m \sum_s L_{ms}}{\sum_k \sum_c L_{kc}^2} \right) L_{ir} \quad \forall i = 1, \ldots, M; \; \forall r = 1, \ldots, R;$$

For instance, applying the above transformation on the previous example, all loadings are scaled to $L_{ir} = 1$ and the output becomes again $G(\mathbf{N}) = 1.7 \cdot 10^{145}$ with a negligible time overhead. However, this form of *static scaling*, compared to the *dynamic scaling* technique of (4), cannot prevent the occurrence of range exceptions and, more seriously, is applicable only to load independent models. More complex scalings for load-dependent models that are have been proposed[16], but anyway do not represent a general solution to numerical instabilities.

Finally, we report another type of overflow problem that should be accounted when solving (2). Consider a PFQN with $R = 2$ classes and assume that a large population vector $\mathbf{n} = (600, 600)$ requires the evaluation of the product-form factor $F_M(\mathbf{n})$. Since $(600 + 600)! \gg (600 + 600)!/600! > 10^{+308}$, the computation of the factorial $F_M(\mathbf{n})$ yields a buffer overflow. A possible solution consists in recursively evaluate $F_M(\mathbf{n})$ from $F_M(\mathbf{n} - \mathbf{1}_s)$ for any class s such that $n_s \neq 0$ using the following relation[22]:

$$F_M(\mathbf{n}) = L_{Ms} c_i^{-1}(\mathbf{n}) F_M(\mathbf{n} - \mathbf{1}_s) \tag{5}$$

Nevertheless, (5) does not guarantee that $F_M(\mathbf{n})$ lies in the allowed floating point range. For instance, consider a model with $M = 2$, $R = 2$ and the loadings $L_{12} = L_{21} = 5$, $L_{11} = 10$ and $L_{22} = 9$. Let the station 1 be a queue-dependent station with $c(\mathbf{n}) = n$ (i.e. a delay station), and let station 2 be a load-independent station. For a population vector $\mathbf{N} = (150, 150)$ it is $F_2(\mathbf{N}) = 8.99 \cdot 10^{+336}$ which generates a buffer overflow.

Summing up, the exact solution of networks including queue-dependent servers with (2) may require also the scaling of possibly thousands of PF factors. Furthermore a study on the mutual interaction between the dynamic scaling of Lam and the numerical range of the PF factors is missing. These problems sensibly limit the ease of implementation and the applicability of (2) on complex models.

8.3.2. Load Dependent Mean Value Analysis (MVA-LD)

The Mean Value Analysis (MVA) algorithm[20] relates the computation of the average performance measures at population $\mathbf{N}$ with the ones computed for the populations $\mathbf{N} - \mathbf{1}_s$. The following MVA formulas allow the exact solution of a load-dependent model and will be referred to as the MVA-LD algorithm:

$$X_r(\mathbf{N}) = \frac{N_r}{\sum_i \sum_{\mathbf{n}=\mathbf{1}_r}^{\mathbf{N}} n_r L_{ir} c^{-1}(n) p_i(\mathbf{n} - \mathbf{1}_r | \mathbf{N} - \mathbf{1}_r)} \quad (6)$$

$$p_i(\mathbf{n}|\mathbf{N}) = \sum_r L_{ir} c^{-1}(n) X_r(\mathbf{N}) p_i(\mathbf{n} - \mathbf{1}_r | \mathbf{N} - \mathbf{1}_r) \quad (7)$$

$$p_i(\mathbf{0}|\mathbf{N}) = 1 - \sum_{\mathbf{0} < \mathbf{n} \leq \mathbf{N}} p_i(\mathbf{n}|\mathbf{N}) \quad (8)$$

where $p_i(\mathbf{n}|\mathbf{N})$ is the probability that server i contains a population $\mathbf{n}$ when $\mathbf{N}$ jobs are present in the network. Compared to the Convolution Algorithm, the MVA-LD shares a similar computational complexity, but it may require less space since only the performance measures for the the populations $\mathbf{N} - \mathbf{1}_s$ must be kept in memory.

In the case of a load-independent model, Eqs. (6)-(8) simplify to

$$X_r(\mathbf{N}) = \frac{N_r}{\sum_i L_{ir}[1 + \sum_s Q_{is}(\mathbf{N} - \mathbf{1}_r)]} \quad (9)$$

$$Q_{ir}(\mathbf{N}) = L_{ir} X_r(\mathbf{N})[1 + Q_{ir}(\mathbf{N} - \mathbf{1}_r)] \quad (10)$$

Note that the algorithm resulting from Eqs. (9)-(10) is numerically robust. Instead, numerical exceptions may affect the MVA-LD algorithm, where

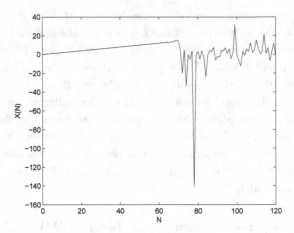

Fig. 8.1. Effects of round-off errors on a single-class throughput $X(\mathbf{N})$ of a load-dependent model.

(8) has been frequently addressed as a source of instability. In particular, if the bottleneck server is a load-dependent server, then the narrow difference between $p_i(\mathbf{0}|\mathbf{N})$ and zero may produce underflow errors and round-offs to negative numbers (note that overflows cannot occur being $p_i(\mathbf{n}|\mathbf{N}) \leq 1$ for all $\mathbf{n}$). These may result on a wrong evaluation of the performance indices as shown in Fig. 8.1 for a single-class throughput $X(\mathbf{N})$ of a load-dependent model. More importantly, round-off errors and underflows cannot be addressed by static or dynamic scaling since queue-lengths are unaffected by such transformations. Reiser[21] have proposed a general variable scaling technique to avoid underflow exceptions. The stabilisation of round-off errors is instead discussed in the next section.

8.4. Improved Stabilisation Techniques

8.4.1. *Software Stabilisation*

As seen in the previous sections, numerical issues of various types may occur during the implementation of queuing-networks algorithms. Despite static and dynamic scaling may be prevent some overflows and underflows, they do not constitute a final solution to the problem. We now discuss a general solution to numerical exceptions in queuing networks algorithms.

Most of the existing algorithms for load-dependent models have been defined during the 70s and 80s. Since then, a number of software libraries for arbitrary precision arithmetics have been developed[2], e.g. the GNU GMP library[11]. These allow to prevent round-off and range exceptions at software level. The basic idea behind arbitrary precision arithmetics is providing library functions for storing and manipulating floating-point numbers with a user-customizable number of precision bits. In this way operations are performed through a software layer and precision is obtained at the expense of an increased computational effort.

Such operations can be handled either with straightforward or state-of-the-art algorithms. If the required precision increase during computation, the latter can limit the overheads. For example, a multiplication between numbers with N precision bits can be performed with the Karatsuba multiplication[12] in $O(N^{1.585})$, much less than the $O(N^2)$ of a traditional multiplication.

It is clear that the major advantage of arbitrary precision arithmetics over the techniques presented in queuing networks literature is the ease of implementation. Scalings and round-offs are automatically handled by the software library. Nevertheless, we can show that the overhead connected to the use of arbitrary arithmetics is almost equivalent to that of dynamic scaling. Table 8.1 and Table 8.2 report CPU times for the Normalisation Constants algorithm for two groups of networks where $M = 2$, $M_Q = 0$ and $M = 2, M_Q = 2$. In all cases, we considered a set of 30 random matrices with $R = 2$, $N_r = N/R$ for all $r = 1, \ldots, R$. Three versions of the Convolution Algorithm have been tested:

- NC does not use any scaling technique for $G(\mathbf{N})$;
- NC-DS implements a dynamic scaling[15] of $G(\mathbf{N})$;
- NC-GMP employs arbitrary precision arithmetics for $G(\mathbf{N})$;

Equation (3) was used to solve the models in which $M_Q = 0$. Since we were interested in timing results, we measured the execution times of NC regardless of the correctness of the results. For the same reason the queue-dependent stations have been set to $c_i(\mathbf{n}) = n$. We considered a maximum timeout of $1500s$ for each algorithm. We point out that the implementations of NC-DS and NC-GMP were obtained with minor modifications from the code of NC, and no additional optimisations were employed.

As we can see from the reported CPU times, the overhead introduced in load independent models (see Table 8.1) by arbitrary precision arithmetics is comparable to that produced by dynamic scaling. However, the implementation of NC-GMP has been straightforward, compared to the time required by the implementation of NC-DS. Furthermore, if we look at the load-dependent models of Table 8.2, we see that the amount of overhead of

Table 8.1. Execution times for unstable and stabilised *load independent* ($M = 2$, $M_Q = 0$ and $N_1 = N_2 = N/2$) multiclass normalisation constant algorithms.

N	1000	2000	3000	4000	5000	6000
NC	0.08s	0.33s	0.74s	1.32s	2.06s	2.95s
NC-DS	0.40s	1.97s	4.82s	8.89s	14.24s	20.75s
NC-GMP	0.50s	1.99s	4.47s	8.16s	12.80s	18.52s

Table 8.2. Execution times for unstable and stabilised *load dependent* ($M = M_Q = 2$ and $N_1 = N_2 = N/2$) multiclass normalisation constant algorithms.

N	50	100	200	300	400	500
NC	0.02s	0.19s	2.58s	17.91s	63.07s	159.38s
NC-DS	0.11s	1.52s	24.71s	127.68s	415.99s	> 1500s
NC-GMP	0.13s	0.41s	7.68s	47.90s	168.47s	459.47s

the dynamic scaling is sensibly bigger than that of NC-GMP. This is easily explained by noting that while performing the convolution (2), an increasing number of rescalings (4) must be applied at each step in NC-DS. Thus, our experimental results show that the software stabilisation approach is more efficient (and in our practice simpler to implement) than dynamic scaling.

8.4.2. *Stabilisation of MVA-LD with Two Customer Classes*

Currently, the best approach for dealing with round-off errors consists in replacing (8) with the generalised Reiser's equation[21]:

$$p_i(\mathbf{0}|\mathbf{N}) = \frac{X_r(\mathbf{N})}{X_r^{-i}(\mathbf{N})} p_i(\mathbf{0}|\mathbf{N} - \mathbf{1}_r) \qquad (11)$$

for each server $i = 1, \ldots, M$ and for any class r such that $N_r \geq 1$. In (11) $X_r^{-i}(\mathbf{N})$ denotes the class-r throughput of the i-complementary system. Note that since MVA-LD requires the evaluation of M_Q probabilities $p_i(\mathbf{0}|\mathbf{N})$, M_Q complementary systems have to be evaluated with (11). Since each of these complementary systems suffer the same instability problems of the original MVA-LD, this would lead to a repeated application of (11) with a combinatorial growth of the number of systems to be solved. Specifically, for a model consisting of $M = M_Q$ queue-dependent servers the number $E(M)$ of complementary systems to be evaluated with (11) is

$$E(M) = \sum_{m=1}^{M} \binom{M}{m} = 2^M - 1$$

which grows exponentially with M. Thus, it is clear that if the number of queue-dependent servers is greater than two or three, then MVA-LD has probably no advantage over (2).

In order to avoid this difficulty, the following solution has recently been proposed[7] for load-dependent models with $R = 2$ customer classes. Define

$$L_{ir}^*(\mathbf{N}) = \sum_{\mathbf{n}} L_{ir} c_i^{-1}(\mathbf{n}) p_i(\mathbf{n} - \mathbf{1}_r | \mathbf{N} - \mathbf{1}_r)$$

for $N_r \geq 1$, and $L_{ir}^*(\mathbf{N}) = L_{ir} c_i^{-1}(\mathbf{n})$ otherwise. Let also

$$\pi_{kr}(\mathbf{N}) = \frac{p_k(\mathbf{0}|\mathbf{N} - \mathbf{1}_r)}{X_r^{-k}(\mathbf{N})}$$

Then, it is possible to show that if $N_1 > 0$ and $N_2 > 0$, then

$$X_1(\mathbf{N}) = \frac{\pi_{k2}(\mathbf{N})}{\pi_{k1}(\mathbf{N}) L_{k2}^*(\mathbf{N}) + \pi_{k2}(\mathbf{N}) L_{k1}^*(\mathbf{N}) + \pi_{k1}(\mathbf{N}) \pi_{k2}(\mathbf{N})} \quad (12)$$

$$X_2(\mathbf{N}) = \frac{\pi_{k1}(\mathbf{N})}{\pi_{k1}(\mathbf{N}) L_{k2}^*(\mathbf{N}) + \pi_{k2}(\mathbf{N}) L_{k1}^*(\mathbf{N}) + \pi_{k1}(\mathbf{N}) \pi_{k2}(\mathbf{N})}$$

$$= \frac{\pi_{k1}(\mathbf{N})}{\pi_{k2}(\mathbf{N})} X_1(\mathbf{N}) \quad (13)$$

For the special case $N_1 = 0$ we have

$$X_1(\mathbf{N}) = 0, \quad X_2(\mathbf{N}) = \frac{1}{L_{k2}^*(\mathbf{N}) + \pi_{k2}(\mathbf{N})} \quad (14)$$

Similarly, when $N_2 = 0$ it is

$$X_2(\mathbf{N}) = 0, \quad X_1(\mathbf{N}) = \frac{1}{L_{k1}^*(\mathbf{N}) + \pi_{k1}(\mathbf{N})} \quad (15)$$

The computational requirements of the algorithm resulting from the above formulas are similar to those of MVA-LD. Nevertheless, the computation is stable with respect to round-off errors without resorting to arbitrary precision libraries. Thus, for reasonable population size, no underflows are generally experienced. The effect of the stabilisation algorithm is shown in Fig. 8.2 for the example considered in the next section.

8.4.2.1. Numerical Example

Let us consider a closed load-dependent queuing network model with $R = 2$ customer classes and with $M = M_Q = 2$ queue-dependent stations. Station loadings are:

$$L_{11} = 10, \, L_{12} = 5$$
$$L_{21} = 5, \, L_{22} = 9$$

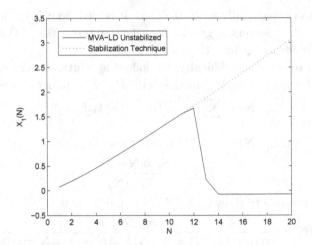

Fig. 8.2. Effect of stabilisation of $X_1(\mathbf{N})$ as $N_1 = N_2 = N$ grow for the system of Sec. 8.4.2.1 with two queue-dependent servers. After the population $\mathbf{N} = (12, 12)$ the MVA-LD algorithm exhibits fatal round-off errors.

while the capacity functions are $c_1(\mathbf{n}) = n^2$ and $c_2(\mathbf{n}) = n$. We show how to apply the algorithm defined in previous section for a simple population vector is $\mathbf{N} = (2, 1)$.

We begin the recursion on the number of stations $m = 1, \ldots, M$. We denote with $\mathcal{M} = \{i, k\}$ a model with two stations labelled by i and k.

Complementary system $\mathcal{M}_1 = \{1\}$.
Note that $\mathcal{M}_1$ is the 2-complementary system. Since only a single station is present, all jobs are in the same resource. Thus, the throughputs are easily computed as

$$X_r^{-2}(\mathbf{N}) = \frac{N_r}{L_{1r}c_1^{-1}(\mathbf{N})N}$$

and for the populations under consideration we get

$$X_1^{-2}(1,0) = 0.100, \quad X_2^{-2}(1,0) = 0.000$$
$$X_1^{-2}(0,1) = 0.000, \quad X_2^{-2}(0,1) = 0.200$$
$$X_1^{-2}(1,1) = 0.200, \quad X_2^{-2}(1,1) = 0.400$$
$$X_1^{-2}(2,1) = 0.600, \quad X_2^{-2}(2,1) = 0.600$$

No other information is kept in memory for $\mathcal{M}_1$.

Complete system $\mathcal{M}_2 = \{1, 2\}$.

Population **n**=(0,0): initialise $p_2(0, 0|0, 0) = 1$.

Population **n**=(1,0): from equations (15) and (7) we get $X_1(1, 0) = 0.067$, $X_2(1, 0) = 0$, $p_2(1, 0|1, 0) = 0.330$. Finally, using Reiser's equation (11) we get $p_2(0, 0|1, 0) = X_1(1, 0)X_1^{-2}(1, 0)p_2(0, 0|0, 0) = 0.670$.

Population **n**=(0,1): similarly to **n**=(1,0), we exploit (14) to compute $X_1(1, 0) = 0$, $X_2(1, 0) = 0.071$, $p_2(0, 1|0, 1) = 0.643$, $p_2(0, 0|0, 1) = X_2(0, 1)X_2^{-2}(0, 1)p_2(0, 0|0, 0) = 0.357$.

Population **n**=(2,0): the computation is analogous to that of **n**=(1,0) and we obtain $X_1(2, 0) = 0.171$, $X_2(2, 0) = 0$, $p_2(2, 0|2, 0) = 0.286$, $p_2(1, 0|2, 0) = 0.143$, $p_2(0, 0|2, 0) = 0.571$.

Population **n**=(1,1): the computation is analogous to that of **n**=(1,0) and we obtain $X_1(2, 0) = 0.171$, $X_2(2, 0) = 0$, $p_2(2, 0|2, 0) = 0.286$, $p_2(1, 0|2, 0) = 0.143$, $p_2(0, 0|2, 0) = 0.571$.

Population **n**=(2,1): using (12) we compute $X_1(2, 1) = 0.182$ and $X_2(2, 1) = 0.086$. From (7) we compute $p_2(0, 1|2, 1) = 0.221$, $p_2(1, 0|2, 1) = 0.123$, $p_2(1, 1|2, 1) = 0.443$, $p_2(2, 0|2, 1) = 0.061$ and $p_2(2, 1|2, 1) = 0.111$. Finally, we can apply Reiser's equation (11) to derive $p_2(0, 0|2, 1)$ either from $p_2(0, 0|1, 1)$ or $p_2(0, 0|2, 0)$. In general, in order to let $p_2(0, 0|\mathbf{n}) \to 0$ smoothly also after one between $p_2(0, 0|\mathbf{n} - \mathbf{1}_r)$ or $p_2(0, 0|\mathbf{n} - \mathbf{1}_s)$ as produced an underflow, it is sufficient to choose the maximum between the two obtained values of $p_2(0, 0|\mathbf{n})$. In this case, since $p_2(0, 0|1, 1)$ and $p_2(0, 0|2, 0)$ are far from the bounds of the representation range, the choice is indifferent and we finally get $p_2(0, 0|2, 1) = 0.041$.

Output: The throughputs are $X_1(2, 1) = 0.182$, $X_2(2, 1) = 0.086$. The queue-length probabilities are available only for station 2. However, in the special case $M = 2$ the probabilities for station 1 are easily computed from the relation

$$p_1(\mathbf{n}|\mathbf{N}) = p_2(\mathbf{N} - \mathbf{n}|\mathbf{N})$$

If other queue-lengths are required, these must be computed using (7) and the computed throughputs of system $\mathcal{M}_2$.

8.5. Conclusions

In this chapter we considered numerical instabilities that arise in computational algorithms for product-form queuing networks. In particular we focused on load-dependent models, showing how the application of arbitrary precision arithmetics can solve the problem more efficiently than previously proposed techniques. Furthermore, we gave an illustrative example of a new technique for avoiding round-off exceptions in load-dependent MVA when only two customer classes are present.

References

1. ANSI/IEEE, "IEEE Standard for Binary Floating Point Arithmetic" Std 754-1985 edition (New York) 1985.
2. Bailey, D.H., "High-Precision Arithmetic in Scientific Computation, Computing in Science and Engineering" *AIP/IEEE Comp. Soc.*, (2005) (to appear).
3. Baskett, F., Chandy, K. M., Muntz, R. R., Palacios, F.G., "Open, Closed, and Mixed Networks of Queues with Different Classes of Customers" *Journal of the ACM*, **22**(2), (1975) 248-260.
4. Berger, A.W., Kogan, Y., "Dimensioning Bandwidth for Elastic Traffic in High-Speed Data Networks" *IEEE Trans. on Networking*, **8**(5), (2000), 643-654.
5. Bruell, S.C., Balbo, G., Afshari, P.V., "Mean Value Analysis of Mixed, Multiple Class BCMP Networks with Load Dependent Service Stations" *Perform. Eval.*, **4**, (1984), 241-260.
6. Buzen, J. P., "Computational algorithms for closed queueing networks with exponential servers" *Commun. ACM*, **16**(9), (1973), 527-531.
7. Casale, G., *The Throughput Analysis of Product-form Queueing Networks* (Ph.D. Thesis, Dip. di Elettronica ed Informazione, Politecnico di Milano, Milan, Italy) 2006 (to appear).
8. Conway, A. E., Georganas, N.D., "RECAL - A New Efficient Algorithm for the Exact Analysis of Multiple-Chain Closed Queueing Networks" *JACM*, **33**(4), (1986), 768-791.
9. Denning, P. J., Buzen, J. P., "The Operational Analysis of Queueing Network Models" *ACM Computing Surveys*,**10**(3), (1978), 225-261.
10. Gordon, W.J., Newell, G.F., "Closed Queueing Systems with Exponential Servers" *Operations Research*, **14**(2), (1967), 254-265.
11. Granlund, T.,"GNU MP: The GNU multiple precision arithmetic library Version 4.1.2" 2003. http://www.swox.com/gmp/.
12. Karatsuba, A., Ofman, Yu, "Multiplication of Many-Digital Numbers by Automatic Computers" Doklady Akad. Nauk SSSR 145, (1962), 293-294. (*Translation in Physics-Doklady* **7**, (1963), 595-596)
13. Kobayashi, H., Gerla, M., "Optimal routing in closed queuing networks" *ACM Transactions on Computer Systems*, **1**(4), (1983), 294-310.
14. Lazowska, J., Zahorjan, J., Graham, G.S., Sevcik, K.C., *Quantitative*

System Performance:Computer System Analysis Using Queueing Network Models (Prentice-Hall) 1984.
15. Lam, S. S.,"Dynamic scaling and growth behavior of queuing network normalization constants" *Journal of the ACM*, **29**(2), (1982), 492-513.
16. Lavenberg, S.S., *Computer Performance Modeling Handbook* (Academic Press, Inc) 1983.
17. Lavenberg, S.S., "A Perspective on Queueing Models of Computer Performance" *Perform. Eval.*, **10**(1), (1989), 53-76.
18. Reiser, M., Kobayashi, H., "Queueing networks with multiple closed chains: Theory and computational algorithms" *IBM J. Res. Dev.*, **19**, (1975), 283-294.
19. Reiser, M., Kobayashi, H., "Numerical methods in separable queueing networks" *Studies Manage Sci.*, **7**, (1977), 113-142.
20. Reiser, M., Lavenberg, S.S., "Mean-value analysis of closed multichain queueing networks" *Journal of the ACM*, **27**(2), (1980), 312-322.
21. Reiser, M., "Mean-value analysis and Convolution Method for Queue-Dependent Servers in Closed Queueing Networks" *Perform. Eval.*, **1**(1), (1981), 7-18.
22. Sauer, C.H., Chandy, K.M., *Computer System Performance Modeling* (Prentice-Hall, Englewoods Cliffs, N.J.) 1981.

CHAPTER 9

Modelling and Simulation of Interdependent Critical Infrastructure: The Road Ahead

Emiliano Casalicchio[1], Paolo Donzelli[2], Roberto Setola[3], and Salvatore Tucci[1,4]

[1]*Dipartimento di Informatica Sistemi e Produzione,*
Università di Roma Tor Vergata, Roma, Italy
E-mail: {casalicchio, tucci}@ing.uniroma2.it

[2]*Computer Science Department,*
University of Maryland-College Park, 20742 MD, USA
E-mail: donzelli@cs.umd.edu

[3]*Complex Systems & Security Lab.,*
Università CAMPUS Bio-Medico, Roma, Italy
E-mail: r.setola@unicampus.it

[4]*Presidenza del Consiglio dei Ministri,*
Ufficio per l'Informatica, Roma, Italy
E-mail: s.tucci@governo.it

Developed societies base their existence on services and resources provided by an increasingly complex network of interdependent critical infrastructures, from power distribution and communication networks, to logistics and healthcare systems, to the Internet. Understanding the behaviour of this system of systems under normal and exceptional circumstances is a mandatory step in order to reduce our dependence and design strategies to increase its dependability. Unfortunately, its complexity overcomes the capability of available methods. In the chapter, we provide an overview of the emerging methodologies and tools, discussing relative potentials and limits.

9.1. Introduction

In contemporary societies, individuals and organisations increasingly depend on availability, reliability, robustness, security, and safety (i.e., dependability) of many technological infrastructures. Ranging from electricity, gas, fuel, and water distribution networks, to transportation, communication, and healthcare systems, to banking and financial networks, they are generally referred to as *critical infrastructures*[1].

Although designed as logically separated systems to deliver different services, over the years these infrastructures have become highly interdependent: each of them relying (directly or indirectly) on the services provided by the others. The large number of interdependencies makes these infrastructures a complex and highly vulnerable system of systems[2]. Indeed, a failure in any of them may easily propagate to the others, with the result of affecting a large, geographically distributed, and largely unforeseeable set of users[3].

Various accidents have demonstrated this scenario. The failure of the telecommunication satellite Galaxy IV in 1998, for example, beside creating communication problems (almost 90% of the pagers were affected), led to significant difficulties also in the transportation system: numerous flights were delayed due to the absence of high-quote weather information, while refuel operations on highways became difficult as gas stations could not process credit card transactions. Another example is provided by the *worm slammer* that rapidly spread worldwide in 2003. It caused problems to financial networks (in US and Canada about 11.000 ATM machines went out-of-service), to air traffic management systems (many flights were delayed or cancelled), to emergency services (the Seattle 911 service became unavailable), and to energy distribution infrastructures (the tele-control systems of two US-utilities firms were affected). Finally, the criticality of these infrastructures has been demonstrated also by the communication blackout occurred in Rome, Italy, at the beginning of 2004. In this case, the air-conditioning failure of a major node of the telecommunication network resulted in a 5-hour black-out of land and wireless telecommunications (affecting almost all the telecom providers), malfunctions of the financial circuits, and problems at the Rome international airport, where 70% of the check-in desks became unavailable.

As well as being highly vulnerable to internal failure, critical infrastructures, due to their increased relevance, might become targets for terrorist and criminal actions. Indeed, attacks could be carried on against infrastructures to create damage, panic, mistrust, or even to increase the effect of other initiatives, e.g., slowing down emergency services and delaying rescue operations in concomitance with more traditional acts of terrorism.

These evidences have contributed to raise the level of attention of governments and international organisations (UN, NATO, EU, etc.) on the need to improve the dependability of these infrastructures[1,4]. The related strategies are usually referred to as Critical Infrastructure Protection (CIP).

In this context, Information & Communication Technologies (ICT) play a double role. In order to improve efficiency and flexibility, ICT are largely adopted at all levels of infrastructures governance (including field and operational control): the massive amount of information to be managed and the tight time constraints overcome the capabilities of human operators and require the adoption of sophisticated ICT solutions. As result, due to their pervasiveness, any ICT failure may have dramatic consequences, as shown, for example, by the 2003 blackout in North-Eastern American: the US-Canada Commission recognised that the failure of some modules of the control systems played a fundamental role in the chain of events that led to the power outage.

Thus, ICT are by themselves the reason for new and unpredictable vulnerabilities and threats that should be taken into account together with the most traditional ones. Recent statistics show a surge of attacks to information systems and a change in their origin: if until 2000 the majority of the attacks have been launched by insiders, in the period 2001-2003 attacks generated by outsiders have become predominant.

In addition to introduce specific vulnerabilities, the spread of ICT, the convergence of communication and information systems, and the increased need for interoperable systems are producing an unprecedented level of interdependency between these infrastructures.

The need to increase the security of the ICT component of critical infrastructures has led to the development of specific strategies, generically indicated as Critical Information Infrastructures Protection (CIIP).

To develop strategies suitable to improve the dependability of these infrastructures we need to be able to understand the global behaviour of this complex *system of systems*. This is a very challenging task, and we are in large part unready because *"the conventional mathematical methodologies that underpin today's modelling, simulation and control paradigms are unable to handle the complexity and interconnectedness of these critical infrastructures"*[5]. In this contest, we need to *"develop creative approaches"*, suitable to be adopted as decision support systems for risk management and resource investment[2].

9.2. Modelling and Simulation of Interdependent Infrastructures

Whereas modelling and simulation techniques of individual infrastructures represent a rather well developed field (and numerous tools are available on the market), techniques for multiple, interdependent infrastructures are still immature.

In the following, we describe some of the approaches currently under development. These studies are primarily devoted to understanding the consequences of a failure element in an infrastructure: for example, which other infrastructures could be affected (through domino or higher order effects), the geographical extension of the phenomenon, the economic impacts, and so on. In particular, current approaches aim to provide insights into infrastructures operations during extreme and rare events, such as major natural disasters or terrorism attacks. Given the rarity of these events, and the great and rapid innovation that characterise the today techno-social scenario, we cannot base our strategies only on record of historical data, but multiple simulations with stochastic variations need to be adopted to obtain useful information on structural characteristics of these events and their impact[2,7]. The goal of these tools is not to portray the exact consequences associated with each single event, but provide useful inputs to recovery plans, reconstruction, and mitigation strategies.

In the literature, there are two main classes of modelling techniques: *Interdependencies Analysis* and *System Analysis*.

Interdependencies analysis techniques are mainly qualitative approaches used to identify critical infrastructures, to discover their interdependencies and to generate macro-analysis scenario. On the other side, System Analysis techniques are simulation-intensive approaches, generally developed using agent-based approaches[6], suitable for more detailed and quantitative studies and able to generate (more or less precise) crisis scenarios.

9.2.1. *Interdependency Analysis*

A simple way to study failure propagation among interdependent infrastructures is to analyse how inoperability, i.e. inability of an element to perform its intended function, is propagated through interconnected infrastructures[8]. Let us denote x^i_{k+1} the inoperability of the i-th infrastructure at time $k+1$; it depends on the level of inoperability of other infrastructures:

$$x^i_{k+1} = \min\left\{f_i(x^1_k, x^2_k, ..., x^n_k) + c_i, 1\right\} \quad 0 \leq x^i \leq 1$$

where the *min* operator formalises the constraint on x^i, and c_i is the risk of inoperability inherent in the i-th infrastructure (due to accidental events, natural disaster or acts of terrorism). Assuming that f_i might be approximated as a linear function, the whole model can be represented as

$$\mathbf{x}_{k+1} = \mathbf{A}\mathbf{x}_k + \mathbf{c} \tag{1}$$

where the entries $\{a_{ij}\}$ of matrix $\mathbf{A}$ represent the degree of incidence of inoperability of j-th infrastructure on the i-th one. Then the overall consequences of a failure is the steady-state solution of (1), i.e., $\bar{\mathbf{x}} = (\mathbf{I} - \mathbf{A})^{-1}\mathbf{c}$.

Notice that this very simple formulation, as illustrated by the following example, is able to emphasise the dangerous consequences of interdependencies. Let us consider four interdependent infrastructures: a Power Plant, a Transportation system, a Hospital, and a Grocery[8]. The impact of a hurricane that destroyed the 50% of the transportation system can be calculated as the steady-state solution of the following system:

$$\mathbf{x}_{k+1} = \begin{pmatrix} 0 & 0.9 & 0 & 0 \\ 0.4 & 0 & 0 & 0 \\ 1 & 0.8 & 0 & 0 \\ 1 & 0.9 & 0 & 0 \end{pmatrix} \mathbf{x}_k + \begin{pmatrix} 0 \\ 0.5 \\ 0 \\ 0 \end{pmatrix}$$

The steady-state solution $\bar{\mathbf{x}} = (0.7\ 0.78\ 1\ 1)^T$ shows that the Hospital and the Grocery are completely inoperable but also, due to interdependencies, that the transportation system is further degraded, reaching the 78% of inoperability.

A similar, but more general approach is represented by the *influence model*[9] where each infrastructure is modelled as a Markov chain whose evolution is influenced not only by its own state, but also by the states of the chain of the neighbouring sites.

Petri nets[10] have been also proposed for the study of failure propagation, but this approach requires the analyst to consider, besides the places representing the different infrastructures, also a large number of phantom places needed to represent propagated failures.

Failure propagation and performance degradation may be investigated via a multi-layer model[11]. In this formulation each layer represents a single infrastructure described in term of nodes (active element), intra-links (functional dependencies inside the infrastructures), and inter-links (functional dependencies with respect to elements belonging to other infrastructures).

As noted by many authors, one of the hardest challenges with this type of model is the definition of the correlation terms that describe infrastructures interdependencies.

Indeed to correctly understand interdependencies existing among the different infrastructures, one has to consider a multitude of "dimensions" in order to take into account the technical, economic, business, social, and political factors that affect infrastructure operations[7]:

Environment. Infrastructures operate in an environment that affects and is influenced by them.

Level of coupling. Systems are tight or loosely linked; their interactions are linear or complex, etc.

Infrastructure characteristics. Infrastructures analysis can be performed on different "scales", depending on the objectives of the analysis, e.g.:

- Geographic scale - infrastructures span physical spaces ranging from cities, to regions, national and international level
- Spatial scale - infrastructure are composed by a hierarchy of elements[12]: *parts* are aggregated in *units* that compose *subsystems*. Subsystems are grouped into *systems* and a collection of systems forms an *infrastructure*.
- Temporal range - Infrastructure dynamics span from milliseconds (e.g., power system operation), to hours (e.g., gas, water, and transportation system operations), to years (e.g., infrastructure upgrades and new capacity).

Type of failure. Interdependence-related disruptions or outages can be classified into three main cases:

- Cascading. A cascading failure occurs when a disruption in one infrastructure causes the failure of a component in a second infrastructure.
- Escalating. An escalating failure occurs when an existing disruption in one infrastructure exacerbates an independent disruption of a second infrastructure.
- Common cause. A common cause failure occurs when two or more infrastructures are disrupted at the same time due to some common event.

Type of interdependency. Linkages among infrastructures can be classified accordingly to their nature:

- Physical Interdependency. Two infrastructures are physically interdependent if functioning of one infrastructure depends on the physical output(s) of the other.

- Cyber Interdependency. An infrastructure presents a Cyber dependency if its state depends on information transmitted through the information infrastructure.
- Geographical Interdependency. A geographic interdependency occurs when elements of multiple infrastructures are in close spatial proximity. In this case, particular events, such as an explosion or a fire could create correlated disturbances or failures in these geographically interdependent infrastructures.
- Logical Interdependency. Two infrastructures are logically interdependent if the state of each depends on the state of the other via control, regulatory or other mechanisms that cannot be considered physical, geographical or cyber.

Cyber interdependency is a relatively new phenomenon strictly related with the pervasiveness of ICT and specifically with the integration of computerised control systems (e.g., Supervision Control and Data Acquisition, SCADA, or Distributed Control Systems, DCS) with other information system, or public communication networks (e.g., Internet). In this way, cyber-interdependency tends to become an absolute and global characteristic of infrastructures, while the other types of interdependency are relative properties. An infrastructure may be physically, logically or geographically interdependent with a specified (generally small) set of infrastructures, while cyber-interdependency potentially couples an infrastructure with every other infrastructure that uses the *cyberspace* and this in spite of their nature, type or geographical location. Incidentally, note that considerations of this kind suggested to many Governments to pay great attention to cyber-threats, and specifically to the vulnerability that SCADA exposed to cyberspace could induce on critical infrastructures[3].

Notice that a different classification of type of interdependencies[13] distinguish them in *direct* and *indirect* links. The first ones are conceived and voluntarily constructed to directly contribute to normal operations, of infrastructure, while *indirect link* does not normally exist, neither specifically constructed nor planned. They are created due a failure and emerge from the modification of the "area" in which infrastructures operate.

The difficulties to integrate into a single model all the elements needed to understand the behaviour of critical infrastructures suggested to some authors to adopt a Hierarchical Holographic Modelling (HHM) approach[14,15]. In HHM, the analyst defines a multitude of (mathematical) models each of them able to clarify and describe some aspects of the problem. Even if each model addresses a different aspect of the systems (components, functions, hierarchical structure, etc.), the overall model offers an acceptable representation of the system. This kind of approach has been used to analyse the transportation system of the state of Virginia in the US and its interdependencies with other infrastructures[16].

Similarly, infrastructures can be modelled as a stack of layers[17]. At the bottom of the stack, there are *Physical Infrastructures*, e.g. roads, pipelines, cables, and electricity generators. The second layer groups the *Operation and Management of Infrastructures*, which encompasses the network control, capacity and routing management. On the top of the stack, there are the *Products and Services of Infrastructures*, which concerns the supply and the use of infrastructure-based products and/or services, such as energy provision or public transportation.

This approach can be further generalised, in order to describe each infrastructure via a three-layer model[18] (see Fig. 9.1):

- *Physical Layer:* The physical component of the infrastructure, e.g., the grid for the electricity network;
- *Cyber Layer:* Hardware and software components of the system devoted to control and manage the infrastructure, e.g., SCADA and DCS;
- *Organisational Layer:* Procedures and functions used to define activities of human operator and to support cooperation among infrastructures.

Within each infrastructure, elements belonging to different layers interact via *intra-dependency links*, while elements of the same layer but belonging to different infrastructure may interact via *inter-dependency links*.

As evident, these approaches are very attractive to emphasise macro-scale consequences of interdependencies allowing stakeholders and experts from different areas to share a common framework.

Unfortunately, due to their oversimplified and abstract nature, they appear to be inadequate to support definition of implementable strategies to improve dependability of critical infrastructures.

9.2.2. System Analysis

Agent-Based Modelling (ABM) is at the moment the most promising technology to model and simulate interdependent infrastructures[6,7,19].

ABM uses a bottom-up modelling approach: the whole model is obtained considering a population of *agents* that autonomously elaborate information and resources in order to produce their outputs. Interactions among

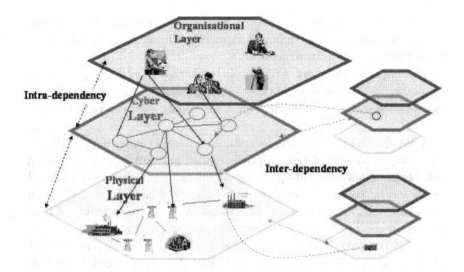

Fig. 9.1. An infrastructure may be decomposed into three layers: Physical, Cyber and Organisational layer. Each element inside to an infrastructure may have relations (dependency), with other elements of the same infrastructure and layer, with elements belonging to the same infrastructure but at different layer (intra-dependency links) or with element belonging to different infrastructures but of the same layer (inter-dependency links).

them results in emerging of behaviours that cannot be predicted by any single agent[7].

In this framework, an agent is an entity with *location*, *capabilities*, and *memory*. The location defines where the agent is in the space (a physical or an abstract space such as Internet). What the agent can perform is defined by its capabilities. An agent can modify its internal data representation (*perception* capability), can modify its environment (*behaviours* capability), can adapt itself to environment's changes (*intelligent reaction* capability), can share knowledge, information and common strategies with other entity (*cooperation* capability), and can execute actions without external intervention (*autonomy* capability). Finally, past experience (for example, overuse or aging) and data defining the entity state represent the agent's memory.

Using these capabilities, ABM allows to capture aggregation phenomena and to provide a natural framework to describe very complex systems.

ABM has been applied by the Argonne national laboratory to build EM-CAS (Electricity Market Complex Adaptive System Model)[a], a simulator

---

[a]www.dis.anl.gov/CEEESA/EMCAS.html

of electricity market. Similarly, ABM is currently used within the Sandia national laboratory for modelling Adaptive Intelligence Aggregation, Economic Systems, and Infrastructures[b].

The implementation of ABM exploits the advantages of agent-based computing[20]. The agent-based programming technology allows production and reuse of simulation libraries: the most widely used are Repast[22] and Swarm[21].

ABM can be implemented also using distributed computing paradigms. For example, the use of Grid, service-oriented and distributed computing technologies allow to build-up the agent infrastructure to run distributed simulation on computer networks, that aggregate enough computational power to simulate the behaviour of systems composed of thousand of hundred of agents.

A different ABM-based formulation for simulation of interdependent infrastructures is given by EPOCHS[24]. EPCHOS is designed to provide high-quality simulations of electric power scenarios while simultaneously modelling the behaviour of computer communications protocols in realistic networks. To this end, three commercial and widely used simulators have been federated into a message-broker framework. In particular, this framework integrates the electromagnetic transient simulator PSCAD/EMTCD, which provides detailed simulations of power systems, with the electromechanical transient simulation engine PSLF, used by energy industries to simulate real-world situations, and with the communication simulator Network Simulator 2 (NS2). NS2 is the most widely used simulator for communication networks based on Internet standard protocols (e.g., TCP/IP). Each of these simulators is considered as an agent that interacts with the others, exchanging results through a proxy agent that works also as time scheduler for the whole simulator.

A similar approach has been proposed to analyse the impact of electric power blackout on transportation and health care systems, federating simulators of the different infrastructures via a Java message broker[18].

One disadvantage of these simulation models is that the complexity of the software programs tends to obscure the underlying assumptions and the inevitable subjective inputs[1].

Moreover, these approaches suffer from the difficulties of acquiring detailed information about each infrastructure. Indeed, infrastructures' owners often consider these data very sensitive and are generally reluctant to make them available.

---

[b]www.cs.sandia.gov/capabilities/AgentBasedModeling/index.html

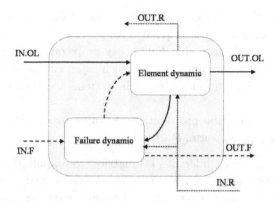

Fig. 9.2. Active element's dynamic proposed in CISIA[23]

In order to partially overcome these drawbacks some authors have proposed CISIA (**C**ritical **I**nfrastructure **S**imulation by **I**nterdependent **A**gents). This is a methodology for evaluate short-term effects, i.e. without take into account fixing activities and plant wearing, of one or more failures in heterogeneous interdependent infrastructures in terms of both failure propagation and performance degradation[19,23].

In CISIA, infrastructures are modelled via highly abstract active elements (agents) which exchange a limited and common set of quantities (messages). The internal behaviour of each agent is described via the interconnected dynamics shown in Fig. 9.2: one models the dynamics associated with the service provided by the agent, while the other considers the evolution of the failures (this is a mix of propagation and internally generate events). CISIA consider five different kinds of dependencies (relations) among agents, each of them is described via an incidence matrix:

An **Operative Incidence Matrix**; where the i-th row represents the set of agents that need the output of the i-th agent to perform their activities;

A **Requirement Incidence Matrix**; where the i-th row represents the set of agents providing the resources needed by the i-th agent.

Three **Fault Incidence Matrices (FIMs)**; where the presence of a 1 in the ij-th position means that a failure may be propagated from the i-th agent to the j-th agent. Notices that in this model the existence of propagation path does not imply straight fault propagation, but the model consider also the nature of the failure and the type of corresponded target agent. Each fault incidence matrix highlight a different type of interdependencies[7]:

- *Physical*: models faults propagated via physical linkages (i.e., those related to exchange of physical quantities);
- *Geographical:* emphasises that faults may propagate among agent that are in close spatial proximity. Events such as an explosion or fire could create correlated disturbances to all the systems localised in the spatial neighbour.
- *Cyber:* used to describe propagation of faults associated with the cyberspace (e.g., virus, worm, etc.).

The authors use three different FIM matrices to simplify the interdependencies' discovery. Indeed, physical interdependencies, and then the possibility of failure propagation across the underlined channels, are, generally, well known to infrastructure's experts and could be read from the functional schemas. On the other side, experts less understand geographical interdependencies, but they can be discovered superimposing infrastructures' maps. Cyber interdependencies are the less understood and the less considered into risk management plans, but, for some aspects, the most important from a security point of view[3] and the most difficult to model too.

9.3. Conclusions

While providing an overview of the emerging methodologies and tools, and discussing relative potentials and limits, in this chapter we have highlighted how modelling and simulation analysis techniques currently available lack the capabilities necessary to represent, study, and analyse critical infrastructures.

New approaches and new synergic combinations of existing approaches need to be explored. In addition, a change of perspective could result necessary in order to extend the currently mainly technical oriented analysis methods to encompass also business and social related issues.

Appendix – National and International Initiatives

The first official act on CIIP was the institution of the President's Commission on Critical Infrastructure Protection in 1996 by President Clinton (for more information see the CIIP Handbook[1]). The work of this commission is at the base of the Presidential Decision Directives 62 and 63 (1998). These directives impose to develop and implement plans to protect government-operated infrastructure, and called for a dialog between the government and the private sector to develop a National Infrastructure Assurance Plan.

These Directives produced great attention on the problem worldwide and many countries established working groups and structures devoted to the definition of strategies to protect their critical information infrastructures[25].

The events of 11th September 2001 impressed new attention on CIIP. In US was established the Department of Homeland Security and on February 2003 two complementary strategic documents were realised:

- The *National Strategy to Secure Cyberspace*[3] which recognises how securing the cyberspace is an extraordinary challenge that requires a co-ordinated effort from the entire society and government. The document defines the cyberspace as an *"interdependent network of information technology infrastructures,"* and depicts cyberspace as the nervous system or control system of the society.
- The *National Strategy for Physical Protection of Critical Infrastructure and Key Assets*[2] devoted to define a strategy to reduce the nation's vulnerability w.r.t. acts of terrorism by protecting the national critical infrastructure and key assets from physical attacks.

Many countries increased the level of awareness on the problem, and created dedicated coordination structures; among others: in Germany was established the *Federal Office for Information Security* (BSI), in Sweden the *Swedish Emergency Management Agency* (SEMA), in the United Kingdom the *National Infrastructure Security Co-ordination Centre* (NISCC), in Canada the *Office of Critical Infrastructure Protection and Emergency Preparedness* (OCIPEP), in Italy the *Working Group on Critical Information Infrastructure Protection* (PIC).

The activities of these structures (and of that instituted in other countries) were mainly devoted to:

- Understand risks related to national critical infrastructures, emphasising their criticalities, their interdependencies, and their vulnerabilities;
- Define strategies to mitigate these vulnerabilities;
- Improve infrastructures' robustness with respect to physical and cyber threats;
- Define emergency, contingency and continuity plans;
- Support R&D on CIIP, and specifically that devoted to better understand interdependent infrastructure's behaviours, and to develop technologies intrinsically robust, secure and safe.

Even international organisations have focused their attention on the problem. The G8 organised in March 2003 the first international conference of G8's CIIP senior experts that released the *G8 eleven principles on CIIP*[26], the 58th General Assembly of United Nation adopted resolution n. 58/199 on *Creation of a global culture of cybersecurity and the protection of critical information infrastructures*[27], the EU Commission released the communication COM(2004)702 on *Critical Infrastructure Protection in the fight against terrorism*[28] and establish *ENISA* (European Network & Information Security Agency), NATO has analysed CIIP in the "Information Operation" considering both civil protection and anti-terrorism[29].

In all these initiatives, R&D represents a corner-store. Indeed, there is no doubt that CIIP will be one of the major R&D challenge in the future. Recent publications and overviews show that R&D in the field of CIP/CIIP is undertaken by a large variety of actors in each country: research institutions, universities, private sector research institutes, etc. Since CIIP issue is largely interdisciplinary, cooperation is a constant element in all the approaches in R&D. Moreover, the inherently trans-national nature of critical infrastructures and the international nature of threats and vulnerabilities (a good example is the big blackout in Italy in September 2003) make the R&D on CIP/CIIP an obvious element for international cooperation.

In this field, the US is definitively the more active player. Their efforts deal with a large variety of issues such as interdependency analyses, threat analysis, vulnerability and risk assessments, system protection and information assurance. Even there are many actors involved in R&D, the degree of coordination is considerable. Among other initiatives, we can cite:

Homeland Security Advanced Research Projects Agency (HSARPA) devoted to jump-start and facilitate early research and development efforts to address critical needs in homeland defence on the scientific and technological front.

Defence Advanced Research Projects Agency (DARPA) within the Department of Defence sponsors "revolutionary, high-payoff research that bridges the gap between fundamental discoveries and their military uses." The DARPA research and development in the CIIP arena is specifically focused on the security and reliability of US military networks.

National Infrastructure Simulation and Analysis Center (NISAC) founded to promote R&D on CIP/CIIP by Los Alamos National Laboratories and Sandia National Laboratories with the cooperation of other research centres as Massachusetts Institute of Technology, Purdue University, Cornell University, Lucent Technologies and Argonne National Laboratory.

Recently EU Commission funding the Co-ordinated Action **CI²RCO** specifically devoted to build up a Europe-wide approach for research and development on Critical Information Infrastructure Protection. Moreover, in the last years, there ware different EU projects related the topic as: DDSI[c], ACIP[d], AMSD[e] and SafeGuard[f].

References

1. Wenger A., Metzger J., Dunn M., Wigert I., "International CIIP Handbook 2004" (ETH, Swiss Federal Institute of Technology, Zurich) 2004, http://www.isn.ethz.ch/crn/publications/publications_crn.cfm?pubid=224
2. "U.S. The National Strategy for The Physical Protection of Critical Infrastructures and Key Assets", 2003, http://www.whitehouse.gov/pcipb/physical.html
3. "U.S. The National Strategy to Secure Cyberspace", 2003, http://www.whitehouse.gov/pcipb
4. Moteff J., Copeland C., Fisher J., "Critical Infrastructures: What Makes an Infrastructure Critical?" Report for Congress RL31556, The Library of Congress, 2003.
5. Amin, M. "Modelling and Control of Complex Interactive Networks" *IEEE Control System Magazine*, (2002), 22-27.
6. Bonabeau E., "Agent-based modelling: methods and techniques for simulating human systems" 2002, 7280-7287. (Proc. National Academy of Sciences of the United States of America, 2002)
7. Rinaldi S., Peerenboom J., Kelly T., "Identifying, Understanding and Analyzing Critical Infrastructure Interdependencies" *IEEE Control Systems Magazine*, (2001), 11-25.
8. Haimes Y., Jiang P., "Leontief-based model of risk in complex interconnected infrastructures" *Journal of Infrastructure Systems*, (2001), 1-12.
9. Asavathiratham S., Leiseutre B., Verghese G., "The Influence Model" *IEEE Control System Magazine*, (2001), 52-64.
10. Gursesli O., Desrochers A., "Modelling Infrastructures Interdependencies Using Petri Nets", 2003, 1506-1512. (IEEE Int. Conf. on Systems, Man and Cybernetics, 2003)
11. Setola R., Ulivi G., "Modelling Interdependent Critical Infrastructure" in *Recent Advances in Intelligent Systems and Signal Processing* Mastorakis, N. E., et al. ed (WSEAS press) 2003, 366-372.
12. Perrow C., "Normal Accidents: Living with High-Risk Technologies" (New York: Basic Books) 1984, 89-100.

---

[c]www.ddsi.org/DDSI-F/main-fs.htm
[d]www.iabg.de/acip/index.html
[e]www.am-sd.org
[f]www.ist-safeguard.org

13. Benoit R., "A method for the study of cascading effects within lifeline networks" *Int. Journal Critical Infrastructures*, **1** (1), (2004), 86-99.
14. Ezell B., Farr J., Wiese I., "Infrastructure Risk Analysis Model" *Int. Journal of Infrastructure Systems*, **6**(3), (2000), 114-117.
15. Haimes Y., *Risk modelling, assessment and management* (Wiley, New York) 1998.
16. Dryden L., Haggerty M., Lane L., Lee C., "Modelling the Impact of Infrastructure Interdependencies on Virginia's Highway Transportation System", 2004, 275-280. (Proc. of the 2004 Systems and Information Engineering Design Symposium, 2004)
17. Thissen W., Hender P., "Critical Infrastructures: Challenges for Systems Engineering", 2003, 2042-2047. (Proc. IEEE Conf. on System, Man and Cybernetics, 2003)
18. Bologna S., Vicoli G., "Interacting Agents Modelling and Simulation of Large Complex Critical Infrastructures Interdependencies" Italy, 2003, 117-122. (Proc. of Italian Society for Computer Simulation Conference - ISCS 2003)
19. Panzieri S., Setola R., Ulivi G., "An agent based simulator for critical interdependent infrastructures", 26-28 October, Grenoble (FR), 2004. (Proc. 2nd Int. Conf. on Critical Infrastructures, 2004)
20. Luck M., McBurney P., Preist C., "A Roadmap for Agent Based Computing" in "Agent Technology: Enabling Next Generation Computing" AgentLink II, 2003. http://www.agentlink.org/roadmap/roadmap.pdf
21. Swarm Delvelopment Group (SDG), "Swarm Simulation System (Swarm)", http://www.swarm.org
22. Repast Organization for Architecture and Development, "Recursive Porous Agent Simulation Toolkit (Repast)", http://repast.sourceforge.net
23. Panzieri S., Setola R., Ulivi G., "An Approach to Model Complex Interdependent Infrastructures" Praha, 2005. (Proc. of 16th IFAC world-congress, 2005)
24. Hopkinson K., Birman K., Giovanini R., Coury D., Wang X., Thorp J., "EPOCHS: Integrated Commercial Off-the-Shelf Software for Agent-Based Electric Power and Communication", 2003, 1158-1166. (Proc. Winter Simulation Conference, 2003)
25. Ritter S., Weber J., "Critical Infrastructure Protection: Survey of worldwide Activities", 2003, (in Proc. of Critical Infrastructure Protection (CIP) Workshop, Frankfurt, 2003).
26. G8 Principles for Protecting Critical Information Infrastructures, March 2003,
   http://www.usdoj.gov/ag/events/g82004/G8_CIIP_Principles.pdf
27. U.N. Resolution n. 58/199, "Creation of a global culture of cybersecurity and the protection of critical information infrastructures", General Assembly, 2003, http://www.un.org/Depts/dhl/resguide/r58.htm
28. E.U. Communication from the Commission to the Council and the European Parliament, "Critical Infrastructure Protection in the fight against terrorism" COM(2004)702, 20th October 2004

29. CiSP Proceedings, Ed. by Stefan Brem, 2004 (EAPC/PfP Workshop on Critical Infrastructure Protection & Civil Emergency Planning: Dependable Structures, Cybersecurity and Common Standards, 2004)

CHAPTER 10

Stochastic Automata Networks and Lumpable Stochastic Bounds: Bounding Availability

J.M. Fourneau[1], B. Plateau[2], I. Sbeity[2] and W.J. Stewart[3]

[1] *PRiSM, Université de Versailles Saint-Quentin,*
45 Av. des Etats Unis, 78000 Versailles, France
E-mail: jmf@prism.uvsq.fr

[2] *Informatique et Distribution, INRIA Rhones Alpes*
51, Av. Jean Kuntzman, 38330 Montbonnot, France
E-mail: Brigitte.Plateau@inria.fr & ihab.sbeity@imag.fr

[3] *Department of Computer Science, NC State University, Raleigh, NC, USA*
E-mail: billy@csc.ncsu.edu

The use of Markov chains to model complex systems is becoming increasingly common in many areas of science and engineering. The underlying transition matrix of the Markov chain can frequently be represented in an extremely compact form – a consequence of the manner in which the matrix is generated. This means that the definition and generation of large-scale Markov models is relatively easy and efficient in both time and memory requirements. The remaining difficulty is that of actually solving the Markov chain and deriving useful performance characteristics from it.

The use of bounding procedures is one approach to alleviating this problem. Such bounds are frequently more easily obtained than the exact solution and provide sufficient information to be of value. In this chapter, we show how to bound certain dependability characteristics such as steady-state and transient availability using an algorithm based on the storage of the Markov chain as a sum of tensor products. Our bounds are based on arguments concerning the stochastic comparison of Markov matrices rather than the usual approach which involves sample-path arguments. The algorithm requires only a small number of vectors of size equal to the number of reachable states of the Markov chain.

10.1. Introduction

Markov chains have long been used to model availability, performability and dependability of complex networks and computer systems. There are two major steps in their application to modelling problems, the first being the derivation of the states of the Markov chain and the transition matrix **P** which describes the transitions among states. The second is the computation of performance measures from this matrix.

The approach of using Stochastic Automata Networks (SAN)[24] is one of several high level formalisms designed to overcome the first problem since it requires a minimum of memory. Other high level formalism such as Petri nets and Stochastic Process Algebra are now also able to store the transition matrix in tensor form[15,20]. Thus, we know how to design and store huge states spaces and transition matrices. However, tensor-based representations do little to help with the second major step of a Markovian analysis, the derivation of performance measures, despite considerable research in recent years[4,16,26,1]. It is quite impossible to obtain analytic solutions for the steady-state or transient distribution of complex Markov chains. Numerical analysis is possible but the memory space requirements and the time complexity are often too great to allow real problems to be solved in a reasonable time period. (see Stewart[25] for an overview of standard numerical procedures for solving Markov chains).

In this chapter, our concern is not in the computation of the stationary or transient distributions per se, but rather in the computation of measures defined on such distributions. In particular, we shall concern ourselves only with obtaining a measure of the steady state availability. However, rather than compute the exact value of availability we shall instead focus on obtaining an upper bound, since in practical applications it is sufficient to know that the availability is larger than the system requirements. In addition, we shall employ a tensor representation of the Markov chain.

Bounding algorithms for steady-state distributions have been successfully used in the past. They are particularly appropriate for availability models because in these models the probability mass is usually concentrated in a small set of states and the algorithms perform well so long as they do not modify the steady-state balance equations on this subset. The first algorithm was proposed by Muntz et al.[23] and was based on theoretical results of Courtois and Semal[11]. This algorithm has since been improved[7,21,22] to reduce the complexity of the computations, improve the accuracy and consider more complex repair service distributions. These algorithms are based on a perturbation of the steady-state balance equations. The approach we adopt in this chapter is rather different. It is based on the stochastic comparison of sample paths of Markov chains but, unlike the usual approach to sample path arguments, our approach is purely algebraic and algorithmic. The comparison we establish is applicable to both transient and steady-state distributions.

Our approach is to develop algorithms that will allow us translate the transition probability matrix of an irreducible, aperiodic Markov chain into a new probability matrix that is a stochastic bound on the original chain and which, in addition, is ordinary *lumpable*. The solution of this lumped chain, which is usually very small and whose solution may be trivially computed, provides the bounds that we seek. The algorithmic aspects of the stochastic comparison of Markov chains have been surveyed recently[17], some algorithms have been developed[19] and a tool implemented[18]. These algorithms are founded on the basis that the transition matrix of the Markov chain is stored in a compact row wise format. A major result of this chapter is to show how related algorithms may be developed which use the much more efficient SAN formalism. Additionally, the algorithms we develop are computationally more efficient and they can be used to obtain bounds on transient measures.

10.2. Stochastic Automata Networks and Stochastic Bounds

Consider an irreducible and aperiodic Markov chain whose stochastic transition probability matrix is given by $\mathbf{P}$. Then there exists a unique solution to the system of equations $\pi = \pi\mathbf{P}$: the row vector π is the stationary distribution of the Markov chain and its ith element denotes the probability that the Markov chain is in state i at statistical equilibrium. An availability measure is defined by separating the states into two classes, UP states and DOWN states. A state said to be UP if the system is UP for that state; otherwise it is DOWN. The steady-state availability is then given by $A = \sum_{i|i \text{ is UP}} \pi(i)$. Notice that if we order the states such that the UP states have indices that are smaller than the indices of the DOWN states, then we can define A as: $A = \sum_i \pi(i) r(i)$ where the reward $r(i)$ is a non-increasing function which is equal to 1 for the first states of the chain (the UP states) and 0 for all the other states. When the stochastic bound is computed, the states must be ordered this way, even if the natural SAN lexicographical ordering does not satisfy this property. This implies a reordering phase during the execution of the algorithm.

10.2.1. SANs and their Tensor Representations

A SAN[27,24] is a set of automata whose dynamic behaviour is governed by a set of *events*. Events are said to be *local* if they provoke a transition in a single automaton, and *synchronising* if they provoke a transition in more than one automaton. A transition that results from a synchronising event is said to be a *synchronised transition*; otherwise it is called a *local transition*. We shall denote the number of states in automaton i by n_i and we shall denote by N the number of automata in the SAN. The behaviour of each automaton, $\mathcal{A}^{(i)}$, for $i = 1, \ldots, N$, is described by a set of square matrices, all of order n_i. In the literature, SANs are usually associated with *continuous-time* Markov chains. The rate at which event transitions occur may be constant or may depend upon the state in which they take place. In this last case they are said to be functional (or state-dependent) transitions. Synchronised transitions may be functional or non-functional.

In the absence of synchronising events and functional transitions, the matrices which describe $\mathcal{A}^{(i)}$ reduce to a single infinitesimal generator matrix, $Q^{(i)}$, and the global Markov chain generator may be written as

$$\mathbf{Q} = \bigoplus_{i=1}^{N} Q^{(i)} = \sum_{i=1}^{N} I_{n_1} \otimes \cdots \otimes I_{n_{i-1}} \otimes Q^{(i)} \otimes I_{n_{i+1}} \otimes \cdots \otimes I_{n_N}. \quad (1)$$

The tensor sum formulation is a direct result of the independence of the automata, and the formulation as a sum of tensor products, a result of the defining property of tensor sums[12].

Now consider the case of SANs which contain synchronising events but no functional transitions and let us denote by $Q_l^{(i)}$, $i = 1, 2, \ldots, N$, the matrix consisting only of the transitions that are local to $\mathcal{A}^{(i)}$. Then, the part of the global infinitesimal generator that consists uniquely of local transitions may be obtained by forming the tensor sum of the matrices $Q_l^{(1)}, Q_l^{(2)}, \ldots, Q_l^{(N)}$. Stochastic automata networks may always be treated by separating out the local transitions, handling these in the usual fashion by means of a tensor sum and then incorporating the sum of two additional tensor products per synchronising event[24]. The first of these two additional tensor products may be thought of as representing the actual synchronising event and its rates, and the second corresponds to an updating of the diagonal elements in the infinitesimal generator to reflect these transitions. Equation (1) becomes

$$\mathbf{Q} = \bigoplus_{i=1}^{N} Q_l^{(i)} + \sum_{e \in \mathcal{ES}} \left(\bigotimes_{i=1}^{N} Q_{e+}^{(i)} + \bigotimes_{i=1}^{N} Q_{e-}^{(i)} \right). \tag{2}$$

Here $\mathcal{ES}$ is the set of synchronising events. This formula is referred to as the *descriptor* of the stochastic automata network[16,24].

10.2.2. Stochastic Bounds: An Algorithmic Presentation.

Stoyan[28] defines a strong stochastic ordering ("st-ordering" for short) by means of the set of non-decreasing functions. Observe that performance measures such as average population size, tail probabilities and so on are non-decreasing functions. Bounds on the distribution imply bounds on performance measures that are non-decreasing functions of the state indices. In our context, the reward that we wish to bound (i.e., $1 - A$) is a non-decreasing function once the state space has been correctly ordered. In the definition given below, the second part concerning discrete random variables is much more convenient for our algebraic formulation and algorithmic setting, as Example 10.2 clearly illustrates.

Definition 10.1:

(a) Let X and Y be random variables taking values on a totally ordered space. Then X is said to be less than Y in the strong stochastic sense, that is, $X <_{st} Y$ iff $E[f(X)] \leq E[f(Y)]$ for all non-decreasing functions f whenever the expectations exist.

(b) If X and Y are random variables taking values on a finite state space $\{1, 2, \ldots, n\}$ and respectively having p and q as probability distribution vectors, then X is said to be less than Y in the strong stochastic sense, that is, $X <_{st} Y$ iff $\sum_{j=k}^{n} p_j \leq \sum_{j=k}^{n} q_j$ for $k = 1, 2, \ldots, n$.

Example 10.2: Let $\alpha = (0.1, 0.3, 0.4, 0.2)$ and $\beta = (0.1, 0.1, 0.5, 0.3)$. It follows then that $\alpha <_{st} \beta$ since:

$$\begin{bmatrix} 0.2 & \leq 0.3 \\ 0.2 + 0.4 & \leq 0.3 + 0.5 \\ 0.2 + 0.4 + 0.3 & \leq 0.3 + 0.5 + 0.1 \end{bmatrix}$$

It has been known for some time that monotonicity[17] and comparability of transition probability matrices yield sufficient conditions for the stochastic comparison of Markov chains and their transient and steady-state distributions. Furthermore, st-monotonicity and st-comparability of matrices may be completely characterised by linear algebraic constraints (which is not true for other types of ordering)[17]. Assuming that **P** is not "st"-monotone, then our goal is to find another matrix **R** such that

$$\begin{cases} \sum_{k=j}^{n} P_{i,k} \leq \sum_{k=j}^{n} R_{i,k} & \forall\, i,j \\ \sum_{k=j}^{n} R_{i,k} \leq \sum_{k=j}^{n} R_{i+1,k} & \forall\, i,j \end{cases} \quad (3)$$

The first set of inequalities states that **P** is stochastically smaller than **R** while the second set shows that **R** is st-monotone. Another small example will help illustrate this

Example 10.3: Consider matrices **P, R1, R2** and **R3**.

$$\mathbf{P} = \begin{bmatrix} 0.2 & 0.2 & 0.1 & 0.3 & 0.2 \\ 0.1 & 0.2 & 0.1 & 0.5 & 0.1 \\ 0.0 & 0.3 & 0.5 & 0.1 & 0.1 \\ 0.1 & 0.2 & 0.4 & 0.3 & 0.0 \\ 0.0 & 0.1 & 0.0 & 0.9 & 0.0 \end{bmatrix} \quad \mathbf{R1} = \begin{bmatrix} 0.1 & 0.3 & 0.1 & 0.3 & 0.2 \\ 0.0 & 0.2 & 0.1 & 0.5 & 0.2 \\ 0.0 & 0.3 & 0.5 & 0.1 & 0.1 \\ 0.1 & 0.2 & 0.4 & 0.3 & 0.0 \\ 0.0 & 0.1 & 0.0 & 0.9 & 0.0 \end{bmatrix}$$

$$\mathbf{R2} = \begin{bmatrix} 0.2 & 0.2 & 0.1 & 0.3 & 0.2 \\ 0.1 & 0.3 & 0.1 & 0.3 & 0.2 \\ 0.0 & 0.1 & 0.4 & 0.3 & 0.2 \\ 0.0 & 0.0 & 0.3 & 0.5 & 0.2 \\ 0.0 & 0.0 & 0.0 & 0.8 & 0.2 \end{bmatrix} \quad \mathbf{R3} = \begin{bmatrix} 0.2 & 0.2 & 0.1 & 0.3 & 0.2 \\ 0.1 & 0.2 & 0.1 & 0.4 & 0.2 \\ 0.0 & 0.3 & 0.1 & 0.4 & 0.2 \\ 0.0 & 0.3 & 0.1 & 0.4 & 0.2 \\ 0.0 & 0.1 & 0.0 & 0.7 & 0.2 \end{bmatrix}$$

Matrix **R1** is an upper bound of **P** but is not st-monotone (the second row is not st-smaller than the third row). **R2** is monotone but is not an upper bound (the second row of **R2** is not an upper bound of the second row of **P**). Finally **R3** is a monotone upper bound of **P**.

In our research, we seek not only to determine matrices **R** that constitute upper bounds on **P** and which are st-monotone but we shall also impose additional restrictions on the structure of **R** in order to facilitate the computation of the bounds. Specifically, we shall insist that the matrix **R** be ordinary lumpable. Fourneau et al.[19] has shown that ordinary lumpability constraints are consistent with the relations specified by (3). Furthermore, we have designed and implemented an algorithm, called LIM-SUB, which constructs a matrix **R** that possesses all these properties. This algorithm requires memory storage for only two vectors of size equal to the size of the number of states in the Markov chain and storage for the lumped matrix which is much much smaller than the original matrix. This lumped matrix is readily solved and the bounds obtained from it may be applied to the original Markov chain. The concept of Markov chains that are lumpable in the ordinary sense has proved valuable in the past. Its definition is now given.

Definition 10.4: (Ordinary lumpability) Let **Q** be the transition probability matrix of an irreducible finite DTMC and let p_k, $k = 1, 2, \ldots, M$ be a partition defined on the states of this Markov chain. Thus, each state of the Markov chain belongs to one, and only one of the so-called *macro-states* p_k. The chain is said to be *ordinary lumpable* with respect to the partition p_k, if and only if, for all states e and f belonging to the same arbitrary macro state p_i, we have $\sum_{j \in p_i} q_{e,j} = \sum_{j \in p_i} q_{f,j}$, for all macro states p_i, $i = 1, 2, \ldots, M$.

LIMSUB computes the lumped matrix column by column. It also requires some additional computations at the boundaries of the blocks. As indicated in Fig. 10.1, two distinct steps are associated with each block. In this table, $first_k$ and $last_k$ are respectively the first and last indices of macro-state k and it is assumed that the states are ordered according to macro state. We stress that this is for illustration purposes only and that the algorithm works with any arbitrary partition.

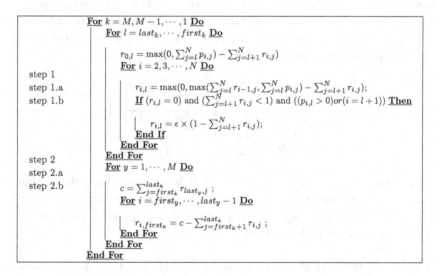

Fig. 10.1. **LIMSUB** Construction of an ordinary lumpable st-monotone upper bounding DTMC **R**.

The first step is based on (3) in which inequalities are replaced by equalities. The relations are unrolled and the equalities are arranged in increasing order for i and in decreasing order for j. Furthermore, to ensure that the irreducibility property is maintained during the bounding process, we avoid deleting transitions and if necessary, we add small sub-block diagonal transitions (step 1.b) that are proportional to a parameter ϵ of order approximately 10^{-7}. The second step modifies the first column of the block in order to satisfy the ordinary lumpability constraint i.e., it assures that each block has the same constant row sum. This constraint is only known at the end of the first step. Thus after the first step, we know how to modify the first column of the block to obtain a constant row sum. Furthermore due to st-monotonicity, we know that the maximal row sum is obtained with the last row of the block (step 2.a). In step 2.b, we modify the first column of the block according to the value of the sum of elements in the last row of the block (variable c in Fig. 10.1). In our algorithmic description, for the sake of simplicity, we use a full matrix representation for $\mathbf{P}$ and $\mathbf{R}$. Naturally, the program uses a sparse matrix representation. Finally, notice that, due to the ordering of the indices, the summations have already been computed and are ready for when we need them. We let them appear as summations to show the relations with the sets of inequalities.

We now illustrate these two steps on the simple example using matrix $\mathbf{P}$ defined above. Assume that the state-space is partitioned into two macrostates: $(1, 2)$ and $(3, 4, 5)$. The matrices below show the first block after the first step (the matrix on the left) and after the second step (the matrix on the right). The values modified during the second step are boxed.

$$\begin{bmatrix} 0.1 & 0.3 & 0.2 \\ 0.1 & 0.4 & 0.2 \\ \hline 0.1 & 0.4 & 0.2 \\ 0.1 & 0.4 & 0.2 \\ 0.0 & 0.7 & 0.2 \end{bmatrix} \quad \begin{bmatrix} \boxed{0.2} & 0.3 & 0.2 \\ 0.1 & 0.4 & 0.2 \\ \hline \boxed{0.3} & 0.4 & 0.2 \\ \boxed{0.3} & 0.4 & 0.2 \\ 0.0 & 0.7 & 0.2 \end{bmatrix}$$

For an arbitrary stochastic matrix $\mathbf{P}$, it has recently been shown that, under easily checkable assumptions on a partition of the state space, the LIMSUB algorithm computes an irreducible and ordinary lumpable st-monotone upper bound matrix $\mathbf{R}$[19].

Theorem 10.5: Let $\mathbf{P}$ be an irreducible DTMC, let $A_1,..., A_p$ be an arbitrary partition of the state space and let ϵ be positive and smaller than 1.0, If the following two conditions are satisfied:

(1) For any n and any state i in macro A_n, there exists a transition $P(i,j)$ with j in A_q and $q < n$ (In other words, there exists a transition to the left at the macro-state level)
(2) For any state i in macro A_1, there exists a transition $P(i,j)$ with j in A_1 and $j < i$ (In other words, there exists a transition to the left within the state level for macro state A_1)

Then, the matrix **R** computed by LIMSUB is a monotone irreducible and ordinary lumpable stochastic upper bound of **P**.

We now show how the algorithm and this theorem establish a common methodology for computing both transient and steady-state bounds.

10.3. LIMSUB and Analysis of Transient Availability

The relevant theorem for the stochastic comparison of Markov chains is now presented[28]. The statement below assumes that **P** the original matrix we want to bound, is not monotone.

Theorem 10.6: Let $\mathbf{X}(t)$ and $\mathbf{Y}(t)$ be two DTMC and **P** and **R** be their respective stochastic matrices. If

- $\mathbf{X}(0) <_{st} \mathbf{Y}(0)$,
- **R** is st-monotone
- $\mathbf{P} <_{st} \mathbf{R}$

Then $\mathbf{X}(t) <_{st} \mathbf{Y}(t)$, for all $t > 0$.

In Fourneau et al.[19], only the steady-state version of this comparison of distributions was considered. However, the theorem states that the sample-paths are ordered. Thus the comparison of distribution is also true for transient distributions and rewards. We know that "st"-bounds are associated with non-decreasing rewards. Then, if $X <_{st} Y$ at time t and $r(i)$ is a non-decreasing reward function, it follows that

$$\sum_i Prob(X(t) = i) r(i) \leq \sum_i Prob(Y(t) = i) r(i)$$

Now consider the partition we have used to build a lumpable bound and let us construct a new reward function. Let A_p be an arbitrary macro-state and define the reward $s(p)$ as the maximum of $r(i)$ for state i in A_p. Clearly:

- $s(p)$ is non-decreasing because the states are initially ordered according to the macro-state.

- for an arbitrary time step t,

$$s(p)Prob(Y(t) \in A_p) \geq \sum_{i \in A_p} r(i)Prob(Y(t) = i)$$

At each moment, the probability of being in macro state p multiplied by the reward $s(p)$ is greater that the sum of the individual rewards multiplied by the probabilities of all the states in macro state p.

As the stochastic matrix associated with Y is lumpable, the left hand-side of the former inequality can be computed using the *lumped* chain. Combining both inequalities we get, for all t,

$$\sum_i r(i)Prob(X(t) = i) \leq \sum_p s(p)Prob(Y(t) \in A_p)$$

Putting everything together, we obtain the following, more general result concerning algorithm LIMSUB.

Theorem 10.7: If the assumptions of Theorem 10.5 are satisfied, if $r()$ are non-decreasing rewards and if the rewards $s()$ at the macro state level are the maximal reward for the individual states, then we can bound the expected rewards $E(r)$ by $E(s)$ at any time t, and the expectation of reward (s) is based on the lumped matrix $\mathbf{R}$.

The algorithm LIMSUB allows us to compute both transient and steady-state rewards and indeed this result is one of the main contributions of this paper. Let us now return to the definition of the transient, cumulative and steady-state availability. Clearly, they are associated with non-decreasing rewards and LIMSUB computes an upper bound for all of them. Naturally, the final numerical solution that is extracted from the lumped matrix depends on the particular problem we want to solve. However, the initial part, i.e., the computation of the bounding matrix, is always the same. The rewards $r(i)$ are equal to 1 if the system is UP in state i and 0 otherwise. The upper bound $s(p)$ is 0 if at least one of the states in macro-state p is 0.

One of the major assumptions of the bounding algorithm is the ordering of the states. First, the rewards must be a non-decreasing function of the state-index. Furthermore, the algorithm is based on a partition of states and assumes that states which belong to the same set have contiguous indices. Finally, it must be clear that the state ordering is a crucial issue for stochastic bounds[14]. The accuracy of the bound is directly influenced by the ordering and hence on the manner in which the bounding matrix is constructed. This implies that we have to reorder the states of the matrix.

Since we represent the transition matrix by a SAN, it is possible to obtain an arbitrary column and to reorder it on the fly. Such an operation is not possible when the matrix is stored on disk.

We now present an overview of algorithm **LIMSUB**. The macro-states are defined by the first and last indices for each macro-state (the vectors **first** and **last** in Algorithm **LIMSUB**). These vectors are built directly from the SAN specification.

```
For k = M, ··· , 1 Do
    For l = last_k, ··· , first_k Do
        colLimsub = GetCol(l, SAN)
        refresh(l, colLimsub, sumP, sumR, firstElementP, endR)
    End For
    normalise(k, endRnorm, endR, firstElementP, sumR, lumpee, sumlumpeeR)
End For
```

Fig. 10.2. Implementation of algorithm **LIMSUB**

The algorithm computes the bounding matrix column per column beginning with the last column. From the presentation of the algorithm **LIMSUB** (see Fig. 10.1), it is clear that it is necessary to store only one column of the matrix **P** at a time. At the boundary of macro-states, a second computation step must be performed. The algorithm is divided into three parts (see Fig. 10.2):

(1) The first step is to obtain a column of the initial matrix from the SAN input. The columns are required in no particular order. Furthermore, as the SAN contains a CTMC and LIMSUB inputs a DTMC, the algorithm must perform an uniformisation procedure. It is known that raw diagonally dominant (RDD) matrices provide more accurate bounds[13]. Thus the uniformisation makes the matrices RDD. In Section 10.4 we describe the function *GetCol* that performs this step.

Since we work on sparse matrices, we store a column of **P** as a linked list using a structure named **ColumnLimsub**. This structure contains 2 fields: one value field, **val**, and a corresponding row index field, **index**.

(2) For the computation part, we require only two vectors, $sumP$ and $sumR$, of size N to represent respectively $sumP[i] = \sum_{k=j}^{n} P_{i,j}$ and $sumR[i] = \sum_{k=j}^{n} R_{i,j}$. The function **refresh()** computes a column of the bounding matrix **R** and updates vectors $sumP$ and $sumR$. An

efficient implementation of this function has been developed and incorporated into the PEPS tool[3] but, for the sake of readability, this is not detailed here. All relevant information concerning an efficient implementation of LIMSUB based on sparse matrix representation for **P** and **R** may be found in the literature[6]. Furthermore, we have established the following property concerning the complexity of the algorithm:

Let x be the number of non-zero elements in an arbitrary column **P** (say i) and let y be the number of non-zero elements in column i of **R**. Then our **refresh** algorithm requires less than $x + y$ steps to compute the ith row of **R**.

There is no simple relation between x and y because the entries in **R** are based not only on the indices of non-zero elements of matrix **P** but also on their values.

(3) For step three, we store the lumped matrix $Rlumpee$ instead of **R**. This requires an array of size $M \times M$ to store the elements of $Rlumpee$ and a vector of size M to compute them: $sumRLumpee$ which contains the sum: $\sum_{k=j}^{r} Rlumpee_{i,j}$. The computation of a column of the lumped matrix is performed by the function **normalise**.

10.4. Algorithm GetCol and its Complexity

Our objective in this section is to show how to extract any individual column in a sparse format and with a ranking of entries coherent with the partition related to the LIMSUB algorithm. The strategy used by Ciardo[10] appears to be an effective method to generate a column from a MDD data structure but requires more memory than the Kronecker approach[8]. The efficiency of this approach comes from avoiding duplicate computation when generating sequential columns. As we have to compute columns according to the partition ordering, this benefit may not materialise. In addition, within the SAN formalism, individual matrices even though they may be relatively small, are stored in a sparse row-wise compact format. An element of any column c is stored using two fields: a value field (val) and a row index field (ind). With functional transitions, the value field contains a function. which prohibits the approach developed for functionless Petri nets[5]. For a sparse matrix **Q**, we shall denote the cth column, by $Q[*, c]$. The nonzero elements within a column, (val_i, ind_i), are ordered according to increasing values of ind_i. Note that val_i is either a positive real or a function of the state space with positive real value and ind_i is a local state index. This may be generalised to elements of the type $\mathcal{S}$ given by $[(val_1, val_2, \cdots, val_d), (ind_1, ind_2, \cdots, ind_d)]$.

We now define an operator on elements of type $\mathcal{S}$, an operator that is a simple concatenation of both subparts.

Definition 10.8: Let $x^1 = (val^1, ind^1)$ and $x^2 = (val^2, ind^2)$ be two elements of type $\mathcal{S}$ where $val^1 = (val^1_1, \cdots, val^1_{d_1})$, $ind^1 = (c^1_1, \cdots, c^1_{d_1})$ and $val^2 = (val^2_1, \cdots, val^2_{d_2})$, $ind^2 = (c^2_1, \cdots, c^2_{d_2})$ and d_1 and d_2 are the dimensions of x^1 and x^2 respectively. The operator $\star$ is defined such that if $X = (VAL, IND) = x^1 \star x^2$ then X is an element of type $\mathcal{S}$ having dimension equal to $d_1 + d_2$ and

$$VAL = (val^1_1, \cdots, val^1_{d_1}, val^2_1, \cdots, val^2_{d_2})$$
$$IND = (c^1_1, \cdots, c^1_{d_1}, c^2_1, \cdots, c^2_{d_2})$$

From this first definition, we now derive a pseudo-tensor form. We shall let $|C|$ denote the length of a vector C.

Definition 10.9: Given an element of type $\mathcal{S}$ and of dimension N (the number of automata of the SAN) $(val_1, \cdots, val_N), (c_1, \cdots, c_N)$, then

$$\textbf{Evaluate}((val_1, \cdots, val_N), (c_1, \cdots, c_N)) = (\prod_{i=1}^{N} val_i(c_1, \cdots, c_N), (c_1, \cdots, c_N)).$$

The term $val_i(c_1, \cdots, c_N)$ denotes the evaluation of the val_i function (which can be a constant function). If C is a vector of elements of type $\mathcal{S}$, **Evaluate**(C) evaluates all its entries.

Definition 10.10: Given vectors $B = \{b^i \mid i = 1, 2, \cdots, |B|, b^i \text{ of type } \mathcal{S}\}$ and $C = \{c^j \mid j = 1, 2, \cdots, |C|, c^j \text{ of type } \mathcal{S}\}$. Then $A = B \star C$, is a vector of length $|A| = |B| \times |C|$ and equal to the vector $b^i \star c^j$, ranked according to the lexicographic ordering on i, j, $i = 1, 2, \cdots, |B|$ and $j = 1, 2, \cdots, |C|$. The cost of this operation is of the order of $|B| \times |C|$.[a]

These definitions will be used to introduce the sparse generalised tensor product of columns of SAN matrices. In this context, it is necessary, at some point, to evaluate the functions.

Let PSS and S respectively denote the SAN potential state space and the reachable state space of the Markov chain. Let $index_{lex}(x)$ denote the lexicographic index of x in the SAN, given by $index_{lex}(x) =$

---
[a] Note the limiting case when one or both vectors have only one entry.

$(x_1 \cdots x_i \cdots x_N)$, where x_i is the index of the local state of automaton i. Thus $index_{lex}(x)$ is an index in PSS and defines an ordering of states. We shall let $index_{PSS}(x)$ denote the integer indexing of states in $1, 2, \ldots, |PSS|$ with the same ordering of states. Then $index_S(x)$ is an integer indexing in $1, 2, \ldots, |S|$ for reachable states in the same ordering. Finally, let $Par = \{p_1, p_2, \ldots, p_M\}$ be the partition of S on which the Markov chain is lumped. This partition is given by the user and depends on the problem. Thus it might not respect the required lexicographic ordering specified previously. Each subset p_i is defined as a function (f_i) with an argument of type $index_{lex}$ (as is the case of all functions in SANs). For this partition, we define a new set of indices, denoted $index_{Par}$, corresponding to the ordering of states employed by algorithm LIMSUB. This ordering ($order_{Par}$) is such that the indices of the elements of subset p_i are contained in the interval $[first_i \ldots last_i]$ where $first_0 = 1$, $last_0 = |p_0|$, $first_i = first_{i-1} + 1$ and $last_i = first_i + |p_i|$, for $i \geq 1$.

As an example, consider a SAN consisting of two automata A_1 and A_2 of size $n_1 = n_2 = 2$. The states of each automaton are denoted 1 and 2 and the potential state space is $PSS = \{1, 2, 3, 4\}$. Assume that $S = \{1, 3, 4\}$, meaning that state 2 is not reachable. Let $Par = \{p_1, p_2\}$ so that $p_1 = \{3, 4\}$ and $p_1 = \{1\}$. Table 10.1 shows the correspondence among indices.

Table 10.1. Correspondence among indices.

$index_{PSS}(x)$	$index_S(x)$	$index_{lex}(x)$	$index_{Par}(x)$
1	1	(1,1)	3
2	–	(1,2)	–
3	2	(2,1)	1
4	3	(2,2)	2

10.4.1. *Computation of a Column*

From (2), **Q** can be seen as the sum of two matrices Q_l and Q_s, where Q_l is the matrix corresponding to local transitions ($\bigoplus_{i=1}^{N} Q_l^{(i)}$) and Q_s is the matrix corresponding to synchronised transitions. A column C of **Q** is the (sparse) sum of two parts C_l (local part) and C_s. Figure 10.3 outlines the different steps that must be performed in obtaining a column.

```
GetCol(index_Par(C),SAN)

index_lex(C)=GetLex(index_Par(C))
C_l = Generate_local(index_lex(C), SAN)      // Generation of C_l
C_s = Generate_sync(index_lex(C), SAN)       // Generation of C_s
C = Sum(C_l, C_s)
colLimsub= sorting(C) // Sort according to order_Par and uniformize
return(colLimsub)
```

Fig. 10.3. Function GetCol($index_{Par}(C)$, SAN)

The index of a column in the entry of function *GetCol* is of type $index_{Par}$, as required by the algorithm LIMSUB, but the algorithm to build C_l and C_s is based on the $index_{lex}$. The solution that we adopt to translate an $index_{Par}$ into an $index_{lex}$ is to define an array (called *SAN2Par*), indexed by $index_{Par}$, which maps into $index_{PSS}$[b]. In this way, $index_{lex}$ can be computed from $index_{PSS}$ using a function $SAN2Lex(index_{PSS}(x))$ — a standard procedure whose cost is N^2 where N is the number of automata. In the array $SAN2Par$, we know the limits of each partition p_i. These are the parameters $first_i$ and $last_i$ defined previously. Thus, $SAN2Par$ can be seen as the union of M subarrays $SAN2Par_i$, and each subarray is sorted in increasing order according to the lexicographic order of the SAN (which is also the order of the $index_{PSS}$ indexing). The different steps of the above algorithms are detailed in the following subsections. In what follows, the lexicographic index of column C is given by $index_{lex}(C) = (c_1, \cdots, c_i, \cdots, c_N)$.

10.4.1.1. *Computation of the Synchronised Part C_s*

The synchronised part of the descriptor presented in (2) is

$$Q_s = \sum_{e \in \mathcal{ES}} \bigotimes_{i=1}^{N}{}_g Q_{e+}^{(i)} + \bigotimes_{i=1}^{N}{}_g Q_{e-}^{(i)}.$$

From this equation, it may be observed that the synchronised part C_s of column C is given as the sum of two parts ($C_s = C_{s^+} + C_{s^-}$) called respectively the positive part and the negative part. The positive part corresponding to a synchronising event e, called C_{e+}, is the tensor product of columns $Q_{e+}^{(i)}[*, c_i]$ of each positive matrix $Q_{e+}^{(i)}$ and hence C_{s^+} is the (sparse) sum of all the C_{e+} ($e \in \mathcal{ES}$). The negative matrices are diagonal

---

[b]It is cheaper to store $index_{PSS}$ than $index_{lex}$.

matrices used for the normalisation of positive matrices. The negative part corresponding to a synchronising event e, called C_{e-}, is a vector containing a single element (the diagonal element of the matrix) which is equal to the product $\prod_{i=1}^{N} Q_{e-}^{(i)}[c_i, c_i]$. Its index is equal to the index of column C. C_{s-} is the sum of the C_{e-} for all $e \in \mathcal{ES}$. The algorithm is presented in Fig. 10.4, and uses the definitions 10.8, 10.9 and 10.10.

Generate_sync($index_{lex}(C)$,**SAN**)

C_{e+}.Empty();C_{e-}.Empty();C_{s+}.Empty();C_{s-}.Empty() // these vectors are of type S
For $e = 1, 2, \ldots, |\mathcal{ES}|$ **Do**
  For $i = 1, 2, \cdots, N$ **Do**

    If $Q_{e+}^{(i)}$ is not identity **Then**

      $C_{e+} = C_{e+} \star Q_{e+}^{(i)}[*, c_i]$
    Else

      $C_{e+} = C_{e+} \star (1, c_i)$
    End If
    $C_{e-} = C_{e-} \star Q_{e-}^{(i)}[c_i, c_i]$
  End For
  Evaluate C_{e+}; Evaluate C_{e-} // Evaluate functional elements
  $C_{s+} = Sum(C_{s+}, C_{e+})$
  $C_{s-} = Sum(C_{s-}, C_{e-})$
End For
$C_s = Sum\ (C_{s+}, C_{s-})$
return(C_s)

Fig. 10.4. **Function Generate_sync** – Generation of synchronised part of the column

The complexity of this algorithm is of order $\sum_{e \in \mathcal{ES}} \times \prod_{i=1}^{N} |Q_{e+}^{(i)}[*, c_i]|$ to which must be added the cost of the K function evaluations.

10.4.1.2. *Computation of the Local Part* C_l

As a tensor sum is a sum of tensor products, the previous algorithm could be used for the local part. However a much simpler scheme can be derived as shown in Fig. 10.5. For each local matrix $Q_l^{(i)}$, this algorithm consists of putting all entries of $Q_l^{(i)}[*, c_i]$, into a particular structure *temp*, by means of function *Change*. To each element (val, b_i) of column $Q_l^{(i)}[*, c_i]$, there corresponds an element $(val, c_1 \cdots c_{i-1} b_i c_{i+1} \cdots c_N)$ in the set *temp*. The elements in this set *temp* are ordered according to the lexicographic order of the SAN. The complexity of this algorithm is $\sum_{i=1}^{N} |Q_l^{(i)}[*, c_i]|$. The cost of evaluating functional transitions must be added to this cost.

```
Generate_local(index_lex(C),SAN)

C_l.Empty()
For i = 1, 2, ··· , N Do
    temp.Empty()
    For each element (val, b_i) of column Q_l^{(i)}[*, c_i] Do
    |   temp.add(val(c_1 ··· c_{i-1} b_i c_{i+1} ··· c_N), Change(index_lex(C),b_i,c_i))
    End For
    C_l = Sum (C_l,temp)
End For
return(C_l)
```

Fig. 10.5. **Function Generate_local** – Generating the local part of a column

10.4.2. *Sorting and Uniformisation*

The procedure *refresh*, used in the main algorithm (Fig. 10.2), requires a column *colLimsub* of type **ColumnLimsub** to be sorted according to $order_{Par}$, where each entry is associated with its index $index_{Par}$. Thus we must transform a list of (entry, $index_{lex}$) into a list (entry, $index_{Par}$) sorted according to $order_{Par}$. This is done in two steps: first, using the function $Lex2SAN(index_{lex}(x))$ (the inverse of function $SAN2Lex(index_{PSS}(x))$ previously used in Subsection 10.4.1) $index_{PSS}$ is computed from $index_{lex}$, and from the $index_{lex}$ of an element, we can also know the subset p_i to which it belongs (using the functions f_i). Secondly, a dichotomic search of $index_{PSS}$ is performed in the subarray $SAN2Par_i$ (as described in Subsection 10.4.1).

Assume that $|p_i| = \frac{|S|}{M}$ ($i = 1 \cdots M$), the complexity of sorting all elements of the columns C according to $order_{Par}$ is $|C| \times (log \frac{|S|}{M} + \mathcal{F} + N^2)$, where $\mathcal{F}$ is the cost of evaluating the functions f_i[c]. We also have to uniformise *colLimsub* before passing it to the procedure *refresh*: all elements are divided by the maximum diagonal element multiplied by 2, which must be nonzero since all states indexed by $index_{Par}$ are reachable.

10.5. Example: A Resource Sharing Model Subject to Failure

In this section we examine the performance of our algorithm when applied to a resource sharing model that is subject to failure. Our interest in this experiment is in obtaining some idea of the closeness of the bounds

---

[c][This cost can be reduced using a MDD to store the reachable state space[9,10]]

generated by the algorithm and in estimating the computation time needed by the algorithm in computing these bounds. In this model, N distinguishable processes share P identical units of a certain resource, where $0 \leq P \leq N$. Each process alternates between a *sleeping* state and a resource *using* state. The number of processes that may simultaneously use the resource is limited to P, so that when a process wishing to move from the *sleeping* state to the active *using* state finds that all P units of resource are occupied, that process instantaneously returns to the *sleeping* state. All probability distribution functions in this model are assumed to be negative exponential distributions. For each process i, $i = 1, 2, \ldots, N$, we shall let $\lambda^{(i)}$ be the rate at which process i awakens from the *sleeping* state wishing to access the resource, and $\mu^{(i)}$ the rate at which this same process releases the resource when it has possession of it.

We add one further property to this system, which to this point is a simple resource sharing model. In addition to the two states *using* and *sleeping*, each process may transit to a third state called *fail*. This state is entered instantaneously from either the *sleeping* or *using* state as a result of a failure signal being received from a synchronising transition. This models the case of a system subject to some global failure, such as an electricity power failure. The rate of occurrence of these failures, i.e., the rate at which this synchronising transition occurs is given by τ_f and its value depends from the number of processes *using* the resource ($\tau_f = (\#using).\alpha$). The rate at which the repair is carried out is denoted by τ_r. Once the failure has been repaired, all processes return to the *sleeping* state to once again begin the resource sharing procedure.

The particular case in which $\lambda^{(i)} = \lambda$ and $\mu^{(i)} = \mu$ for all $i = 1, 2, \ldots, N$ can be shown to be strongly aggregatable[2], and this model becomes trivial to analyse. However, if we assume that $\lambda^{(i)} = \lambda + i\delta$, $i = 1, 2, \ldots, N$, the model cannot be aggregated and requires much more computation time to obtain its solution. Furthermore, since the failure rate is a functional rate, no product-form solution is available for this model. The specific parameter values used in these experiments are given by $\lambda = \mu = 5.0$; , $\tau_r = 1.0$ and the constant of the failure rate equal to $\alpha = 10^{-3}$, for a model with $N = P = 10$. The number of partitions is taken to be equal to $P + 2$ and is defined so that partition k, $k = 0, 1, \ldots, P$ contains all states in which the number of processes in the active *using* state is equal to k; the final partition consists of a single state, the *failure* state. To judge the effectiveness of our bounding approach, we compared it with the exact results for different values of δ_i. Since the exact results cannot be obtained for large models, our comparisons were conducted on models with the relatively small reachable state space of 1,024 states. For values of $\delta = 0.001$, 0.01, 0.1 and 1, we found the exact value and our computed bounds to be respectively, $(4.93E-03, 5.02E-03)$, $(5.10E-03, 5.18E-03)$, $(5.98E-03, 6.45E-03)$ and $(8.32E-03, 9.05E-03)$. The time to compute the bound in all four cases was 0.44 seconds. In summary, the algorithm performs extremely well.

Acknowledgments

This work is supported by ACI Sure-Paths and, in part, by the NSF under research grant number ACI 0203971

References

1. Benoit, A., Plateau, B., and Stewart, W.J., "Memory Efficient Iterative Methods for Stochastic Automata Networks" Technical report INRIA n. 4259, France, (Sept. 2001).
2. Benoit, A., Brenner, L., Fernandes, P. and Plateau, B., "Aggregation of Stochastic Automata Networks with Replicas" *Linear Algebra and its Applications* **386**, (2005), 111-136.
3. Benoit, A., Brenner, L., Fernandes, P., Plateau, B. and Stewart, W.J., "The PEPS Software Tool" (Springer-Verlag) LNCS 2794, 2003, 145-166 (4th International Conference on Modeling Techniques and Tools for Computer Performance Evaluation (Urbana, Illinois, USA), September 2003).
4. Buchholz, P., "An aggregation\disaggregation algorithm for stochastic automata networks" *Probability in the Engineering and Informational Sciences*, **11**, (1997), 229-253.
5. Buchholz, P., Ciardo, G., Donatelli, S. and Kemper, P., "Complexity of Kronecker Operations on Sparse Matrices with Applications to the Solution of Markov Models" (NASA/CR-97-206274 ICASE Report No. 97-66, December 1997).
6. Busic, A. and Fourneau, J.M., "Bounds for point and steady-state availability: an algorithmic approach based on lumpability and stochastic ordering" (Springer-Verlag) LNCS 3670, 2005, 94-108, (2nd European Performance Engineering Workshop, France 2005).
7. Carrasco, J.A., "Bounding steady-state availability models with group repair and phase type repair distributions" *Performance Evaluation*, **35**, (1999), 193-204.
8. Ciardo, G., Forno, M., Grieco, P., and Miner, A, "Comparing implicit representations of large CTMCs" A.N. Langville and W.J. Stewart eds, 2003, 323-327, (Fourth International Conference on the Numerical Solution of Markov Chains NSMC'03 (Urbana, Illinois, USA), September 2003).
9. Ciardo G. and Miner, A.S., "Storage alternatives for larger structured state spaces" (Springer-Verlag) LNCS 1245, 1997, 44-57 (Proc. 9th Int. Conf.

on Modelling Techniques and Tools for Computer Performance Evaluation, 1997).
10. Ciardo, G. and Miner, A., "A data structure for the efficient Kronecker solution of GSPNs" IEEE Comp. Soc. Press, Buchholz, P. ed., 1999, 22-31 (Proc. 8th Int. Workshop on Petri Nets and Performance Models (PNPM'99), Sept. 1999).
11. Courtois, P.J. and Semal, P., "Bounds for the positive eigenvectors of nonnegative matrices and their approximations" *J. ACM*, **31**(4), (1984), 804-825.
12. Davio, M., "Kronecker products and shuffle algebra" *IEEE Transactions on Computers*, **30**(2), (1981), 116-125.
13. Dayar, T., Fourneau, J.M., Pekergin, N., "Transforming stochastic matrices for stochastic comparison with the st-order" *RAIRO-RO*, **37**, (2003), 85-97.
14. Dayar T., Pekergin, N.: "Stochastic comparison, reorderings, and nearly completely decomposable Markov chains" Plateau, B. and Stewart, W. eds. (Prensas universitarias de Zaragoza) 1999, 228-246.(Proceedings of the International Conference on the Numerical Solution of Markov Chains (NSMC'99), 1999).
15. Donatelli, S., "Superposed generalized stochastic Petri nets: definition and efficient solution" (Springer-Verlag) LNCS 815, 1994, 258-277 (Proc. 15th Int. Conf. on Application and Theory of Petri Nets, Zaragoza, Spain, June 1994).
16. Fernandes, P., Plateau, B., Stewart, W.J., "Efficient descriptor-vector multiplications in stochastic automata networks" *JACM*, **45**, (1998), 381-414.
17. Fourneau, J.M., Pekergin, N., "An algorithmic approach to stochastic bounds" (Springer Verlag) LNCS 2459, 2002, 64-88 (Performance evaluation of complex systems: Techniques and Tools, 2002).
18. Fourneau, J.M., Lecoz, M., Pekergin, N. and Quessette, F., "An open tool to compute stochastic bounds on steady-state distributiuon and rewards", 2003, 219-224 (IEEE Mascots 2003).
19. Fourneau, J.M., Lecoz, M., Quessette, F., "Algorithms for an irreducible and lumpable strong stochastic bound" *Linear Algebra and its Applications* **386**, (2005), 167-185.
20. Hillston, J., Kloul, L., "An Efficient Kronecker Representation for PEPA Models" (Springer Verlag) LNCS 2165, 2001, 120-135 (Proc. of the Joint International Workshop on Process Algebra and Probabilistic Methods, Performance Modeling and Verification, Germany, Sept. 2001).
21. Lui, J.C.S., Muntz, R., "Computing bounds on steady state availability of repairable computer systems" *J. ACM*, **41**(4), (1994), 676-707.
22. Mahevas, S., Rubino, G., "Bounding asymptotic dependability and performance measures", 1996, 176-186 (Second IEEE International Performance and Dependability Symposium, USA, 1996).
23. Muntz, R., de Souza e Silva, E., Goyal, A., "Bounding availability of repairable computer systems" *IEEE Trans. on Computer*, **38**(12), (1989), 1714-1723.
24. Plateau, B., "On the stochastic structure of parallelism and synchronization models for distributed algorithms" SIGMETRICS, Texas, 1985, 147-154

(Proc. of the SIGMETRICS Conference, Texas, 1985).
25. Stewart, W.J., *Introduction to the Numerical Solution of Markov Chains* (Princeton University Press) 1994.
26. Stewart, W.J., Atif, K., Plateau, B., "The numerical solution of stochastic automata networks" *European Journal of Operational Research*, **86**, (1995), 503-525.
27. Stewart, W.J. and Plateau, B., "Stochastic Automata Networks for Dependabillity Modelling" 2000 (IEEE Aerospace Conference, USA, 2000).
28. Stoyan, D., *Comparison Methods for Queues and Other Stochastic Models* (Wiley) 1983.

CHAPTER 11

Aggregation Methods for Cross-Layer Simulations

Monique Becker, Vincent Gauthier
GET/INT, Samovar Lab., Evry, France
E-mail: {Monique.Becker, Vincent.Gauthier}@int-evry.fr

André-Luc Beylot, Riadh Dhaou
INPT/ENSEEIHT, IRIT Lab., Toulouse, France
E-mail: {Andre-Luc.Beylot, Riadh.Dhaou}@enseeiht.fr

The behaviour of ad hoc networks is very difficult to predict due to its dynamic characteristics. In this chapter we assess the relevance of including inter-layer interactions to analyse quality of service (QoS) issues in wireless networks so that upper layer(s) processes could take into account the performance of the lower layer when, for example, selecting paths which provide some QoS guarantees. These interactions introduce time-level-space-scale dependencies. Moreover, mobility of such systems make an experimental approach, based on measurements and testing, time consuming. Hence, we advocated that analytic or simulation techniques could be developed instead. Aggregation techniques have been widely proposed to solve large scaled and complex systems and in this chapter we study their suitability in a cross-layer simulator environment.

11.1. Introduction

The performance of wireless systems is extremely sensitive to nodes' mobility, and to the dynamic nature of the traffic and environment. In the recent past the concept of a layered network stack has been at the core of development of networks and standard protocols. However, with the emergence of ad hoc and wireless networks, it has become apparent that the performance of a layered stack approach can be compromised by the specificity of the wireless networks, their interconnection with wired networks and early attempts to just adapt previously proposed quality of service (QoS) mechanisms to wireless environments.

Recently cross layer based approaches have been suggested as a new way of studying and assessing novel QoS mechanisms in heterogeneous network. These mechanisms do not consist of adding extra features like e.g. bandwidth reservation mechanisms or any other ad hoc mechanism, but attempt to adapt the currently available mechanisms at one layer of the stack to its underlying layers. Cross layer concept implies sharing pertinent information between not necessarily adjacent layers with the aim of optimising a global objective, as opposed to the traditional multiple-layer optimisations.

Performance evaluation and optimisation of the whole system is a crucial step in the process of design and validation of these emerging systems.[1] Decomposition has long been recognised as a powerful tool for the analysis of large and complex systems where models of large, complex and dynamic systems can often be reduced to smaller sub-models, for easier analysis, using aggregation/dis-aggregation (decomposition) techniques. However before applying these techniques, the system under analysis should possess some key underlying near-decomposable characteristics[2].

In this chapter we assess the usefulness of including aggregation/dis-aggregation techniques when simulating complex wireless communication systems. We implement these techniques in a simulation environment which is driven by dynamic data and entail the ability to incorporate additional data (either archived or collected online) so that the simulator tool would be able to steer the evaluation process dynamically.

In the first part of this chapter, we review aggregation techniques previously suggested to solve complex Markov chains, queuing networks and physical systems. In the second part, we present cross layer interaction case studies, using the developed dynamic simulation tool.

11.2. Aggregation Methods

The underlying idea of aggregation/dis-aggregation techniques is that large complex systems may share some features of nearly/totally decomposable systems[2]. If this is the case, a large scale and complex system may be divided into subsystems which are easier and faster to solve. Aggregation techniques are in the more basic a two step procedure.

Aggregation step: the solution of each subsystem is derived independently from the other subsystems, or if there are relations between subsystems, they are simplified.

Dis-aggregation step: the global solution is derived from the solutions of the previous step.

Aggregation may be exact or approximate. When it is approximate, the two steps may be used iteratively until convergence[3]. Some examples of communication systems which could resort to aggregation techniques are: An example of space type aggregation is the analysis of a GSM system. In these systems, a base-station controller coverage is an aggregate of base tranceiver coverage and switching centre coverage is an aggregate of base station controller coverage. An example on when time scale aggregation techniques could be of great value is in the analysis of ATM systems[4-6]. The analysis of an ATM network at call level and at cell level simultaneously using, for example, Markov chain models can become very time consuming. The underlying problems lies in the time-scale difference between the two stochastic processes involved.

11.3. Aggregation of Markov Chains

11.3.1. *Introduction*

In the case of Markov chain models, several methods have been proposed in order to benefit from the structural features of the graph associated with the underlying model transition matrix (or infinitesimal generator)[7].

11.3.2. *Decomposability Method of Kemeny and Snell: Theory*

Kemeny and Snell[8] aggregation method is based on the property known as decomposability (or lumpability). In this case, the state space is partitioned into disjoined subsets called aggregates.

$$E = \bigcup_{n \in \Gamma} A_n, \quad \forall\, (i,j) \in \Gamma^2,\ i \neq j,\ A_i \cap A_j = \phi \tag{1}$$

where, Γ is finished or countable. In this framework a system is decomposable if the sum of the rates of transitions from a state of an aggregate A_i towards the states A_j of an aggregate is the same one for all the states of the aggregate A_i.

$$\forall\, (i,j) \in \Gamma^2,\quad i \neq j,\ \forall\, (e_i, f_i) \in A_i^2,\quad \sum_{e_j \in A_j} q_{e_i, e_j} = \sum_{e_j \in A_j} q_{f_i, e_j} \triangleq q^*_{A_i, A_j}. \tag{2}$$

The method of Kemeny and Snell then makes it easy to determine the stationary probabilities of the various aggregates (i.e. the sum of stationary probabilities of the states of an aggregate). For that, let us consider the process $\{Y(t), t \geq 0\}$ with space of states $F = \{A_k\}$ defined in the following way: $Y(t) = A_k \to X(t) \in A_k$. Hence, $\{Y(t), t \geq 0\}$ is then a Markov process with an infinitesimal generator $\mathbf{Q}^* = (q^*_{A_i,A_j})$. And we can obtain

$$\Pr(Y(t+dt) = A_j | Y(t) = A_i) = \Pr(X(t+dt) \in A_j | X(t) \in A_i)$$
$$= q^*_{A_i,A_j} dt + o(dt). \qquad (3)$$

Then the distribution $\pi^* = (\pi^*_{A_k})$ where $\pi^*_{A_k} = \sum_{e_k \in A_k} \pi_{e_k}$ is the solution of:

$$\pi^* \mathbf{Q}^* = 0, \quad \sum_{n \in \Gamma} \pi^*_{A_n} = 1. \qquad (4)$$

Finally, the distribution π^* verifies the system (4). Note that this method determines in a relatively simple way, the stationary probabilities of the various aggregates, but, we do not obtain the probabilities of the states inside an aggregate.

11.3.3. *Example of the Method of Kemeny and Snell*

The proposed model consists of studying a switch of packets with two input ports and two output ports (see Fig. 11.1). The packets are sorted upon arrival according to their destination into two virtual sub-queues. Let us note Q_1 and Q_2 the two queues and Q_{11}, Q_{12}, Q_{21} and Q_{22} the four virtual sub-queues. Q_{ij} is the set of packets arriving on input i and served by the output port j server. The ingress queues are continuously polled in parallel (separately for the two outputs). When a queue is not empty, a packet is

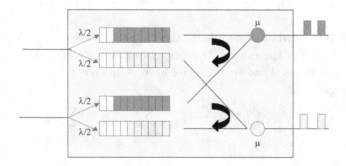

Fig. 11.1. Packets switch with ingress queues.

emptied; if the queue is empty, the other queue is polled. The switching time between queues is negligible (it's a two queue polling system with a null switching time between queues and between cycles).

The arrivals on the input queues are assumed to be Poisson with rate λ on each of the two entries and we assume that the traffic is randomly equitably distributed on the two output ports. So the input traffic into each of the virtual sub-queue is $\lambda/2$. The length of the packets is supposed to be exponentially distributed with mean length $1/\mu$. In this case, we can separately analyse the behaviour of the two output ports. Let us analyse what occurs on the first exit. Let us note $N_1(t)$ the number of packets in virtual sub-queue Q_{11}, $N_2(t)$ the number of packets in virtual sub-queue Q_{12} and $K(t)$ the number of the port which is actually polled. Let us note that when a queue is empty the virtual sub-queues have no real meaning. The polling operation has a meaning only if the system is not empty. There is no switching time, so if a packet arrives to an empty queue it is immediately transferred. Then the process $\{N_1(t), N_2(t), K(t), t \geq 0\}$ can be described by a Markov chain. Let N, $1 \leq N \leq +\infty$ be the capacity of the buffers. Then the Markov chain state space is:

$$E = \{(i,j,k), 1 \leq i \leq N, 1 \leq j \leq N, 1 \leq k \leq 2\}$$
$$\cup \{(0,j,2), 1 \leq j \leq N\} \cup \{(i,0,1), 1 \leq i \leq N\} \cup \{(0,0,0)\}. \quad (5)$$

The decomposition method suggested identifies states aggregates in which there are M packets in the two queues. Note that the method is exact in the two limit cases $N = 1$ and $N = +\infty$. For $N = +\infty$, the system is equivalent to a M/M/1 queue where the discipline of service is changed. It is FIFO on each queue, but not necessarily overall, if a packet arrives at a queue which is relatively highly loaded it may be overtaken by packets which would arrive at the other queue while it waits. Globally, insofar as discipline is work-conserving, and the number of packets waiting for an emission on the first exit (the sum of the numbers of packets waiting at Q_{11} and Q_{21}) constitutes a Markov process with a total arrival rate λ and a service rate μ. We obtain exactly the same graph as that obtained for an M/M/1 queue.

Let us verify it from the transition rates:

— the aggregate $A_0 = \{(0,0,0)\}$. It is a singleton, hence there is no need to verify any property. With a rate $\lambda/2$, there are transitions towards the states $(1,0,1)$ and $(0,1,2)$ which belong to the aggregate A_1. We obtain: $q^*_{A_0,A_1} = \lambda$;

— generally, starting from any state (i,j,k) of an aggregate A_{i+j} there are

transitions, with a rate $\lambda/2$, towards the states $(i+1,j,k)$ and $(i,j+1,k)$ both of which belong to the aggregate A_{i+j+1}. Then, $q^*_{A_n,A_{n+1}} = \lambda$;
— for the departure rates, let us verify that $q^*_{A_n,A_{n-1}} = \mu$. For that, we have to distinguish in each aggregate A_n, $n \geq 0$, the states where one of the queues is empty, from the states where the two queues are not empty;
— from states $(0,n,2)$ and $(n,0,1)$ one will go towards the states $(0,n-1,2)$ and with a rate μ. These states belong to the aggregate A_{n-1};
— from states $(k,n-k,2)$ and $(k,n-k,1)$ where $0 < k < n$, there is a transition towards the states $(k,n-k-1,1)$ and $(k-1,n-k,2)$ with a rate μ. These states belong to the aggregate A_{n-1}.

The property (2) is thus verified and the aggregated graph shown in Fig. 11.2 is obtained:

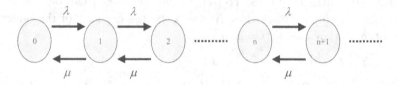

Fig. 11.2. Aggregate process, infinite length queues.

Note that this aggregation method does not permit us to obtain neither the marginal probabilities of occupation of each buffer, nor the probabilities of each state of the initial process.

For the other limit case (buffer size equal to one) the transition diagrams shown in Fig. 11.3 are then obtained.

Finally, the method of Kemeny and Snell makes it possible to obtain the probabilities in stationary regime of each aggregate. Since the system is balanced, the expected number of packets in each queue and the loss probability may then be derived.

11.3.4. *Decomposition Method of Courtois: Theory*

The decomposition method suggested by Courtois is an approximated method for which an error bound can be calculated. A system is known as completely decomposable if variables within constituent sub-systems do not relate to variables of other sub-systems. Almost completely (or nearly) decomposable systems have been referred as systems in which the relation

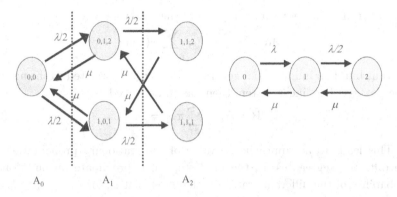

Fig. 11.3. Initial process and aggregated process, queues with size 1.

between the sub-systems' variables are weak[9]. In the case of a Markov process, we will thus say that it is decomposable if the infinitesimal generator may be put in the block diagonal form. In a Markov chain model framework these systems are not interesting since it means that the underlying process is not ergodic (as the associated graph is not strongly connected). Nearly decomposable systems on the other hand are hence more interesting. A Markov process is nearly completely decomposable if its associated infinitesimal generator can be put in the form:

$$\mathbf{Q} = \mathbf{Q}^* + \varepsilon \mathbf{C} \qquad (6)$$

where $\mathbf{Q}^*$ is completely decomposable and ε very small.

The method of Courtois is carried out then in two steps known as aggregation and dis-aggregation.

Step No. 1: Aggregation

The first step consists of determining the conditional stationary probabilities for each aggregate in isolation. Hence for $i \in \Gamma$ the following linear systems is then solved:

$$\pi^*_{A_i} \mathbf{Q}^*_{A_i} = 0, \quad \sum_{e_i \in A_i} \pi^*_{e_i} = 1. \qquad (7)$$

Step No. 2: Dis-aggregation

The second step consists of determining the stationary probabilities of the different aggregates. For that, we determine in an approximate way the rates of transition between aggregates in the following way:

$$r_{A_i,A_j} \approx \sum_{e_i \in A_i} \pi^*_{e_i} \sum_{e_j \in A_j} q_{e_i,e_j}. \tag{8}$$

We thus build an infinitesimal generator $\mathbf{R} = (r_{A_i,A_j})$ related to the aggregate process. The linear system then has to be solved

$$\pi \mathbf{R} = 0, \quad \sum_{i \in \Gamma} \pi_{A_i} = 1. \tag{9}$$

This leads to an approximate value of the stationary probabilities of the different aggregates. In the final step an approximate unconditional probability of the different states is calculated that is, for $i \in \Gamma$, $e_i \in A_i$:

$$\pi_{e_i} \approx \pi^*_{e_i} \pi_{A_i}. \tag{10}$$

This method determines in an approximated way the unconditional stationary probabilities of all states. Courtois also showed that the incurred error was of the order of $O(\varepsilon)$.

11.3.5. *Example of the Decomposition Method of Courtois*

Let us consider the Markov process characterised by the graph shown in Fig. 11.4.

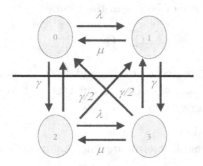

Fig. 11.4. Initial process.

If we assume that γ is comparatively small in magnitude when compared to λ and μ, then the infinitesimal generator:

$$\mathbf{Q} = \begin{pmatrix} -(\lambda+\gamma) & \lambda & \gamma & 0 \\ \mu & -(\mu+\gamma) & 0 & \gamma \\ \gamma/2 & \gamma/2 & -(\lambda+\gamma) & \lambda \\ \gamma/2 & \gamma/2 & \mu & -(\mu+\gamma) \end{pmatrix} \tag{11}$$

can be reduced to

$$\mathbf{Q}^* = \begin{pmatrix} -\lambda & \lambda & 0 & 0 \\ \mu & -\mu & 0 & 0 \\ 0 & 0 & -\lambda & \lambda \\ 0 & 0 & \mu & -\mu \end{pmatrix} \quad (12)$$

which leads at the dis-aggregation step to:

$$\begin{cases} \pi_0^* = \dfrac{\mu}{\lambda+\mu} \\ \pi_1^* = \dfrac{\lambda}{\lambda+\mu} \end{cases} \begin{cases} \pi_2^* = \dfrac{\mu}{\lambda+\mu} \\ \pi_3^* = \dfrac{\lambda}{\lambda+\mu} \end{cases}. \quad (13)$$

Note that in this case the step of determination of the rates of transition between the aggregates is exact (that is, we could have applied the method of Kemeny and Snell):

$$\begin{cases} r_{1,2} = \gamma \pi_0^* + \gamma \pi_1^* = \gamma \\ r_{2,1} = \gamma \pi_2^* + \gamma \pi_3^* = \gamma \end{cases}. \quad (14)$$

From (14), we get:

$$\mathbf{R} = \begin{pmatrix} -\gamma & \gamma \\ \gamma & -\gamma \end{pmatrix} \Rightarrow \pi_1 = \pi_2 = \frac{1}{2} \quad (15)$$

and finally while applying (15):

$$\begin{cases} \pi_0 \approx \dfrac{1}{2}\dfrac{\mu}{\lambda+\mu} \\ \pi_1 \approx \dfrac{1}{2}\dfrac{\lambda}{\lambda+\mu} \end{cases} \begin{cases} \pi_2 \approx \dfrac{1}{2}\dfrac{\mu}{\lambda+\mu} \\ \pi_3 \approx \dfrac{1}{2}\dfrac{\lambda}{\lambda+\mu} \end{cases} \quad (16)$$

11.4. Aggregation of Physical Sub-System

The aggregation methods presented in previous sections are based on mathematical and structural properties of the underlying models. In this section, we present an example on how decomposition can appear naturally when analysing communication systems. Let us consider a GSM cellular network[10,11]. In a GSM cellular network, users are allowed to move and the network providers' aim is to ensure the continuity of intercellular communication throughout the duration of the call. If the movement patterns are analysed (see Fig. 11.5), it can be noted that movement follows certain specific patterns and hence, it may be possible to regard groups of cells as systems aggregates.

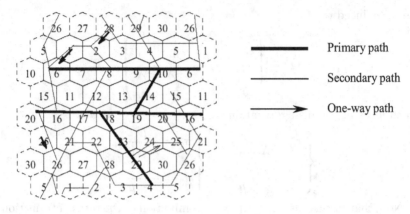

Fig. 11.5. GSM Network example.

When a user goes from one cell to another cell, there is a handoff. When there are many handoffs between two cells, it will be suitable to aggregate them.

In Ref. 4 a network composed of m cells and N mobiles moving between the various cells is considered. The assumptions are the following:

— the residence time in a cell i follows an exponential distribution with holding time $1/\mu_i$;
— the probability that a mobile leaving out cell i and going towards cell j is given by p_{ij}.

The objective of the work reported in Ref. 10 is to determine the average numbers of mobiles in each cell N_i by supposing that there is no blocking in any cell (there are sufficient resources there). In this case, it may be sufficient to determine for each one of the N users the steady state probability π_i that he is in cell i. This can be obtain directly by resorting to the relation $N_i = \pi_i N$.

The total residence time rate is then $\mu = \sum_i \mu_i$ and the a_{ij} elements of the transition matrix $\mathbf{A}$ are:

$$\begin{cases} a_{jj} = 1 - \dfrac{\mu_j}{\mu} \\ a_{ij} = \dfrac{\mu_i p_{ij}}{\mu} \quad i \neq j \end{cases} \tag{17}$$

which has a steady state solution given by the linear system:

$$\pi \mathbf{A} = \pi, \quad \sum_{i=1}^{m} \pi_i = 1. \tag{18}$$

From the physical characteristic of the system under analysis and empirical evidence gathered so far, it appears that the matrix $\mathbf{A}$ is very sparse and hence aggregation techniques may be employed to obtain an approximated solution by aggregating cells between which the probabilities of handoff are high. For a 79 cells configuration, covering France, several aggregates were examined successively. One aggregate was Paris region. The other aggregates were corresponding to large areas. Two methods of resolution were then used: Courtois method and a numerical solution of a block diagonal system by power method[12].

In Ref. 11 we considered this time that the number of users varies in each cell and that the network operator has a limited number of resources in each cell (recall that if the access method used is F-TDMA, time slots are allocated to frequencies). Consequently, it will be necessary to take into account new call arrivals, intercellular call transfers, ends of communications and potential failures. Using the notation already introduced, we also need to consider the following parameters:

— duration of communications follows a negative exponential distribution with mean holding time $1/\gamma$;
— the new connection arrivals is a Poisson process with rate λ_i;
— a cell can accept up to C_i simultaneous connections.

Even though other solutions are possible[11], in this example, we present the simplest version which does not use priority mechanisms between the various types of communications (new calls, intercellular transfers). We then try to determine:

— the blocking probabilities of new calls in each cell: τ_i;
— the blocking probabilities of transfers of calls between the various cells: $\tau_{i,j}$.

Under these assumptions, the vector $\{N(t) = (N_1(t), \ldots, N_m(t)), t \geq 0\}$ corresponding to the number of calls in each cell constitutes a continuous time Markov process whose state space size Ω is equal to $\Pi_{i=1}^{m}(C_i + 1)$.

Note that if traffic is asymmetric and communications are routed between unbalanced cells, the near-decomposable feature that the previous model transition matrix possesses is lost and the methods here discussed may not be valid.

The proposed decomposition technique for the simple case focuses on one cell at the same time. To make the analysis more tractable, it is assumed that the intercellular traffic is Poisson with rate $\lambda_{i,j}$. If this is the case, we can resort to the underlying property of Poisson traffics which makes it possible to equalise the probabilities of blocking of the new calls and of the intercellular transfers: $\tau_i = \tau_{j,i}$.

We obtain, then, a very simple model of a cell which behaves exactly like a $M/M/C_i/C_i$ queue with arrival rate $\alpha_i = \lambda_i + \sum_{j \neq i} \lambda_{j,i}$ and service rate $\nu_i = \mu_i + \gamma$. We then obtain a load $\rho_i = \alpha_i/\nu_i$ for which the probability of blocking equals:

$$\tau_i = \tau_{j,i} = \frac{\frac{\rho_i^{C_i}}{C_i!}}{\sum_{j=0}^{C_i} \frac{\rho_i^j}{j!}}. \tag{19}$$

The problem is that in this formula, the intercellular transfer rates are not known because of potential blocking due to lack of resources. We could however calculate these rates according to the blocking probabilities and the characteristic parameters of the system:

$$\lambda_{i,k} = (\lambda_i(1-\tau_i) + \sum_j \lambda_{j,i}(1-\tau_{j,i}))\frac{\gamma}{\nu_i}p_{i,k}$$

$$= \left(\lambda_i + \sum_j \lambda_{j,i}\right)\frac{\gamma}{\nu_i}p_{i,k}(1-\tau_i). \tag{20}$$

An iterative solution in which we apply successively the aggregation step (19) followed by the dis-aggregation step (20) is envisaged. Note that we can start from an initial solution in which all blocking probabilities are set to zero. In this case the resolution of the linear system (20) is equal to the determination of the average number of passages in each cell if there is no blocking. In this case (20) changes into:

$$\lambda_{i,k} = \left(\lambda_i + \sum_j \lambda_{j,i}\right)\frac{\gamma}{\nu_i}p_{i,k}. \tag{21}$$

11.5. Time-Space Aggregation

There are phenomena for which events at many scales of length or time do make contributions of comparable importance, or at least have a non negligible influence on each other. Hence, any evaluation framework that describes these phenomena must take into account these multi scale contributions.

A typical example of multiple-time-layer-space-scale dependence behaviour is provided by ad hoc[13] and sensor systems[14,15]. When analysing these systems, the choice of granularity depends on the nature of the study. In some cases several levels of modelling of the same set of functions may be necessary, and multiple correlated processes may be identified. The identification of these dependencies and the way in which they should be treated will then become mandatory.

The system is composed of a set of sub-systems distributed in the space (see Fig. 11.6). Each sub-system (e.g. mobile node and/or sensor) is composed of one or more layers (but not necessarily all the layers). There is a cyclic dependence[16] between the various layers $((1, i), (2, i) \ldots$ and $(N, i))$ of each sub-system. This cyclic dependence occurs when several parts of the sub-system are correlated (for example, the performance of the radio channel depends on the traffic while the traffic itself depends on the performance of the radio channel). This dependence can also be characterised between two layers which are not directly adjacent (for example for sub-system i

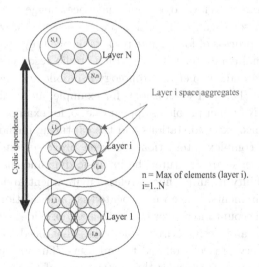

Fig. 11.6. Space and layer aggregations.

there may be a dependence between layers 7 and 1 $(7, i)$ and $(1, i)$ in the case of cross layer systems[17]. Indeed, the cross layer simulator should allow the consultation — in real time — of information available on any network layer. This will enhance the simulator capabilities making it more sensitive to the specificity of the support and mechanisms used on each layer and find optimal conditions considering the whole system instead of optimising each layer locally.

Within each layer of the system, it is possible to aggregate the subsystems. In this example, we refer to space aggregation which may rely on the physical distance between elements but which can also take into account the intrinsic characteristics of each layer. For example, if we consider a cellular system aggregates may correspond to cells. But, generally, aggregates depend on the layers. The aggregates obtained on the layer 2 and on the layer 7 may be different from each other because the dependence (e.g. distance and/or time) between the different sub-systems on those layers may be different. Finally note that by increasing the number of elements of the system, we do not have any guarantee to preserve the space aggregates.

11.6. Layer Aggregation

Telecom networks were designed as layered systems. In this context, protocols were planned to operate on only one layer. However, for some emergent networks, like for example, ad hoc networks, networks operation is becoming more complex and cross layer operations have been suggested. Ad hoc wireless networks are self organizing multi-hop wireless networks where all nodes take part in the process of forwarding packets, hence they are very different from conventional computer networks. First, radio resources are rare and time varying. Second, the network topology is mobile and the connectivity is unpredictable. Third, if we consider, for example, an architecture-based on 802.11 WLAN further problems appear due to, for example, the presence of hidden stations, exposed stations and the capturing phenomena.

Due to the complex interactions between processes and protocols in a traditional layered architecture, the intrinsic dynamic characteristics of resource availability in an ad hoc wireless environment and the emerging end-users' requirements, hence we advocate that a simulation environment that takes into account inter-layer interaction can provide a suitable framework to analyse and design wireless networks that need to maintain the quality of the service (QoS) offered to end-users. The proposed approach allows upper layer(s) to consider the performance of lower layers in order to, for example, select a path which provides QoS guarantees.

11.6.1. Dynamic Simulations

If inter-layer interactions are readily available feedback between layers implicitly creates a controlled set of interactions to enable a particular layer to work with the rest of the stack. This feedback is not easy to implement in a network simulator because each layer works sequentially with it own time-scale, but with a feedback mechanism each layer may not be working sequentially and might have a different time-scale. For example the time scale of a simple 801.11 session is different from a time-scale of a TCP session. To avoid this problem when two entities share the same variable but update or read this value at different time-scales, it is necessary to work with time averaging value. The slowest layer will determine if the layer stack is stable or not[18]. Hence when simulating a network which takes into account a feedback mechanism we need to have a dynamic simulator which creates dynamically feedback information by itself.

11.6.2. Example of Inter-Layer Design

This section reports on the inter-layer interaction between the physical layer and the routing layer that we have implemented in our proposed simulator tool. Our approach differs from Raisinghani's[19] in that our simulator takes into account the time-scale differences between layers. With the structure design proposed by Raisinghani[19] there is no time information stored in the DME (Data Management Entity) and we must work with a time average value, but in many cases we need to work with values which are context aware and time aware (it is not enough to know values it may be necessary to know the context and the time). Any piece of information needs to have a context: for example, at MAC layer, each piece of information is related to a link, but in the routing layer, each piece of information is related to a path, if we want to exchange information between layers, we need also to exchange the context (exchange information about the layer knowledge example: routing information, session information). In our case, for an interaction (feedback information) between the physical layer and the routing layer, routing cannot be aggregated into an average because each piece of information associated to each path has its own context and it own time-scale. We have developed another approach to avoid the time-scale problem and context problem. In this implementation of cross-layer design, the inter-layer information is exchanged through the same path and at the same time as a normal packet through each layer. In each layer a new component is implemented to store and schedule the information which comes from the lower layer to the upper layer (DME). Each piece of information is synchronised with the packet flow through the network stack. This framework permits us to create an inter-layer interaction (feedback relation between layers) between the physical layer and the routing layer without loss of context information and also allows the right synchronisation with the rest of the data flow. Each piece of information can now be linked with routing layer information without any complex synchronisation mechanism. The inter-layer interaction flow (flow of feedback information) is storing and forwarding along the parallel way of the usual data flow (flow of packets in the network stack). The implementations of this inter-layer interaction have been made with JIST and SWANS[20], a Java event driven network simulator.

11.6.3. *Network Protocols which use Inter-Layer Interactions*

Ad hoc networks require adaptive routing schemes to deal with the frequent topology changes and end-user demand requirement. In this section we study a routing protocol which uses inter-layer information to select the route: a trade-off between the number of hops, the theoretical available bandwidth and the stability of the route (it is stable if there is no change about availability of the route when time is running). This inter-layer information includes the signal strength and Signal to Noise Ratio (SNR) available at the MAC layer of an individual host. A new network metric is defined such that a trade-off between the number of hops and the SNR of each individual route is considered. The new metric, available at the network layer, allows us to have a global overview of the best available path. In this protocol, a host initiates route discovery on-demand by broadcasting a route-search packet which propagates to the destination, allowing the destination node to choose a route and return a route reply. In this section we focus on the implementation of the ad hoc On Demand Distance Vector (AODV) routing protocol that takes into account cross layer information to obtain better routing decisions. To do so, we use the SNR of each node to determine the route which will have the best SNR along the available paths. For this purpose we have added the probability of packet drop based on SNR information, in each Route REQuest packet (RREQ packet). Each node that forwards these packets adds its own SNR information, thus updating the SNR of each link along the route. When the destination node receives the RREQ packet, it directly obtains the information about the quality of the route. The destination node then determines the best route and replies by sending a RREP packet so that each node on the route can save the relevant QoS information in its routing table. This route decision mechanism takes advantage of the cross layer information and hence we should expect the quality of the route and the network throughput to be improved. The decision criteria takes into account frame error rate (FER), bit error rate (BER) and signal to interference ration (SIR) information as stated in Eqs. (22)-(24).

$$FER_l = 1 - (1 - BER_l)^n, \qquad (22)$$

$$BER_l = \hat{\alpha}_M Q\left(\sqrt{\hat{\beta}_M\, SIR_l(P)}\right), \qquad (23)$$

$$SIR_l(P) = \frac{P_l G_{ll}}{N_l + \sum_{k \neq l}^{N} P_l G_{lk}}, \qquad (24)$$

where;
n : Number of bits in a packets,
G_{lk} : Path losses from the transmitter on the logical link l to the receiver on the logical link k,
P_l : Transmission Power from the transmitter on the logical link l,
N_l : Background noise on the logical link l,
FER_l : Frame error rate on the logical link l,
BER_l : Bits error rate on the logical link l.

Equations (22)-(24) consider a wireless network with N nodes, and an established logical topology, where some nodes are sources of transmission, and a sequence of connected links $l \in L(s)$ determines a route originating from source s[21]. $\hat{\alpha}_M = \alpha_M/\log_2(M)$ where α_M is the number of nearest neighbours to constellation at the minimum distance, and $\hat{\beta}_M$ in $\hat{\beta}_M = (\log_2 M)/\beta_M$ is a constant that relates minimum distance to average symbol energy, β_M and α_M are constants depending on the modulation and the required bit error rate is given by BER. The BER is derived from the signal to interference ratio for link l (SIR_l) for a given set of path losses G_{lk} (from the transmitter on the logical link k to the receiver on the logical link l) and a given noise N_l for the receiver on the logical link l.

Each node computes the BER and the FER of each input and each forward flow l as a function of the frame size n in bits and BER and SIR

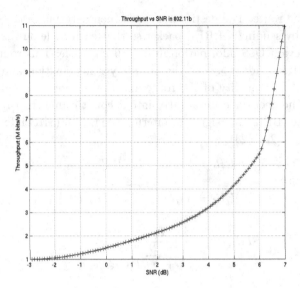

Fig. 11.7. Throughput as a function of the SNR.

which will become an inter-layer information. Hence, the probability of frame dropping on the logical link $l \in L(s)$ could be readily available for the routing protocol.

In our simulations we have modelled the AODV protocol to carry information concerning the probability of frame drop over the path. During the route discovery phase the AODV protocol broadcasts a route-request packet looking for the destination. In our proposed solutions, each node along the path between the source and the destination piggy-backs (in the route-request packets) information about the probability of frame drop. As a result the destination node calculates the probability of a frame drop for the chosen path, P,

$$P = \prod_{l \in P(l)} FER_l, \qquad (25)$$

where l is a sequence of connected links belonging to path $P(l)$. The results obtained from the cross-layer simulations are promising and a highlight of these follows. Let us first show results that prove that it is adequate to design cross-layer models. On Fig. 11.7 the throughput between two nodes with a single hop is presented as a function of the SNR. Figure 11.8 shows the throughput of a two nodes single hop also for a multi-rate protocol like 802.11b.

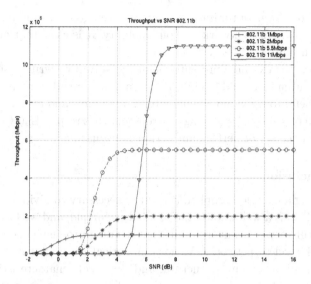

Fig. 11.8. Throughput as a function of the SNR for various data rate (1, 2, 5.5, 11 Mbps).

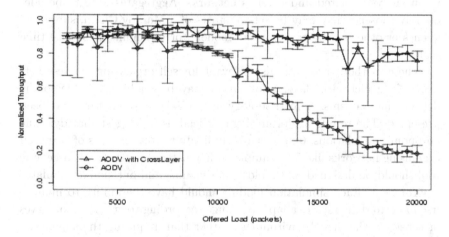

Fig. 11.9. Performance comparison for models with and without cross-layer optimisation.

These results show that in this type of scenario it is advisable to select the route considering SNR due to its observed impacts on the available throughput. Figure 11.9 presents the normalised throughput (throughput/load) as a function of the offered load for the case of AODV without and with cross-layer optimisation, from an average of 30 simulations per points for a 50 nodes network without mobility. It is clear that cross-layer optimisation improves the performance.

Note that the example presented here was solved by writing simulation modules for each layer and having them communicate adequately. In fact it would be useful to use a more generic tool: a dynamic and autonomic simulator, i.e. an on demand operating environment that responds automatically to change in the set of performance criteria evaluations.

11.7. Conclusion

When solving a complex problem it is often necessary to divide it into sub-problems. This is a well known and studied technique, and we have showed that it is still very relevant in solving outstanding problems like routing in ad hoc and wireless networks. Various numerical methods have been proposed to take into account structural and time scale characteristics of the underlying models and they are normally referred to as aggregation methods. We have presented a few of these methods and assess their applicability when analysing wired and wireless networks. Aggregates have to be adequately chosen. They may be sets of states of a Markov Chain, subsets of queues or other sub-systems, or sub-models corresponding to several time scales.

These methods are even also relevant for self organising wireless networks for which cross-layer information may be readily available to the delivery mechanisms. A methodology for cross-layer simulations has been presented. The results are promising and lead us to suggest that dynamic autonomic simulations are a serious candidate to solve models of large self organising networks like for example ad hoc networks. Therefore, a generic tool should be designed. It should be a dynamic and autonomic simulator environment. Each simulation module should have the ability to manage itself and to dynamically adapt to changes according to criteria it observes or senses in the model environment rather than requiring the simulation designer to initiate the changes.

References

1. Gelenbe, E., Pujolle, G., *Introduction to Queueing Networks*, (Wiley), 1987.
2. Courtois, P. J., "On the time and space decomposition of complex structures", *Communication of the ACM*, **28**(6) 1985, 590-603.
3. Dudkin, L. M., Rabinnovich, H., and Vakhutinsky, I., "Iterative aggregation – A new approach to the solution of large scale problems", *Econometrica*, **47**(4), 1979, 821-842.

4. Roberts, J.,"A service system with heterogeneous user requirements", in "Performance of data comm. systems and their applications", G. Pujolle, Ed., (North Holland), 1981, 423-431.
5. Ziram, A., *Aspects multiservices des réseaux large bande: application à l'étude des performances au niveau appel des commutateurs ATM*, (Ph.D. Thesis in Computer Science, U. Pierre et Marie Curie, Paris 6), 1998.
6. Ziram, A., Beylot, A.-L., and Becker, M., "Using an Aggregation Method for the Multiple Resource Sharing Problem in a Multi-service Environment", 1998, 176-182 (6th International Conference on Telecom Systems, Nashville, USA).
7. Moulki, M., Beylot, A.-L., Truffet, L., and Becker, M., "An aggregation technique to evaluate the performance of a two-stage buffered ATM switch", *Annals of Operations Research*, **79**, (1998), 373-392.
8. Kemely, J. G., Snell, J.L., *Finite Markov Chains*, (Van Nostrand-Reinhold, Princeton), 1960.
9. Courtois, P. J., *Decomposability: Queuing and computer system applications*, ACM Monograph Series, (Academic Press), 1977.
10. Mazel, C., *Evaluation des performances par simulations: Application aux canaux de signalisation de systèmes radiotéléphoniques*, (Thèse de docteur-ingénieur de l'Institut National Polytechnique de Grenoble), 1988.
11. Boumerdassi, S. and Beylot, A.-L., "Adaptive Channel Allocation for Wireless PCN", *ACM Journal on Special Topics in Mobile Networks and Applications*, MONET, 4(2), (1999), 111-116.
12. Stewart, W. J., "Numerical methods for computing stationary distributions of finite irreducible Markov chains", in "Advances in computational probability" (Chap. 4), W. Grassman, Ed., (Kluwer Academic Press), 2000, 81-112.
13. MANETs, *Mobile Ad hoc NETwork Group*, http://www.ietf.org/html.charters/manet-charter.html.
14. Xbow, http://www.xbow.com.
15. Zigbee Alliance, http://www.zigbee.org/.
16. Becker, M., Beylot, A.-L., and Dhaou, R., "Aggregation Methods for Performance Evaluation of Computer and Communication Networks", in "Performance Evaluation – Stories and Perspectives", Ed. Gabriele Kotsis, (Oesterreichische Computer Gesellschaft), 2003, 215-230.
17. Dhaou, R., Gauthier, V., Becker, M., and Beylot, A.-L., "Cross Layer Simulation: Application to Performance Modelling of Networks Composed of MANETs and Satellites", 2004, T11.1-T11.30, (2nd Conference HetNet'04, Ilkley, U.K.).
18. Ljung, L., "Analysis of Recursive Stochastic Algorithm", *IEEE Trans. Auto. Control*, **AC-22**, (1977), 551-575.
19. Raisinghani, V.T. and Iyer, S., "Cross-Layer Design Optimisation in Wireless Protocol Stacks", *Computer Communication*, **27**,(2003), 720-725.
20. JIST/SWANS, *Java in Simulation Time Scalable Wireless Adhoc Network Simulator*, http://jist.ece.cornell.edu/.
21. Goldsmith, A., *Wireless Communications*, (Cambridge University Press), 2004.

Modelling of Emerging Networks

CHAPTER 12

Space and Time Capacity in Dense Mobile Ad Hoc Networks

Philippe Jacquet

INRIA and Ecole Polytechnique
91128 Palaiseau Cedex, France
E-mail: philippe.jacquet@inria.fr

In this paper we revisit the two main paradoxes of mobile ad hoc networks: the space capacity paradox and the time capacity paradox. Gupta and Kumar have shown that density increases wireless capacity. Grossglauser and Tse have shown that mobility also increases capacity. We quantify more precisely this properties by setting up the general laws that propagation path must satisfy in presence of traffic flow density. Introducing time constraint in packet delivery, we generalise to a space-time problem with mobile networks.

12.1. Introduction

Mobile ad hoc networks involve nodes that are moving on a network domain and communicate via radio means. The domain of a network can be indifferently a battlefield, a urban quarter, a building floor, etc. The mobile ad hoc networks bring two fundamental paradoxical properties:

- The space capacity paradox;
- the time capacity paradox.

The illuminating result of Gupta and Kumar[2] states that the maximum capacity per node in a flat domain is in $O(\sqrt{\frac{1}{N\log N}})$, N being the network size. Multiplying by N we get the total capacity of a wireless ad hoc network which is in $O(\sqrt{N/\log N})$ and therefore increases in N. This result is in sharp contrast with what can be obtained in a wired network. Assume N nodes connected by wire to the same communication resource (*e.g.* a cable headend). The resource share per node is in average $O(\frac{1}{N})$ and therefore the total capacity remains bounded when N increases. In other words, wireless networks create capacity when the density increases. Therefore adding space to a multihop wireless network increases the capacity: this is the space capacity paradox.

Figure 12.1 displays an example of the space capacity gain in an uniform traffic model described in Ref. 7 for Fisheye-OLSR. OLSR is a proactive link state protocol for mobile ad hoc networks[6], and Fisheye-OLSR is an adaptation of OLSR protocol to very large networks.

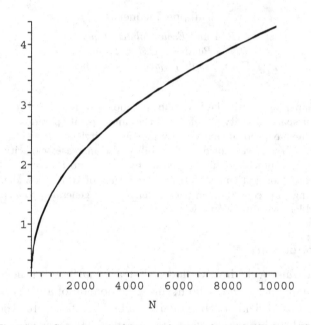

Fig. 12.1. Capacity gain factor with network size for Fisheye-OLSR.

When nodes randomly move, it turns to be more advantageous to store packets for a while on mobile routers instead of forwarding them immediately like hot potatoes. When the mobile router moves closer to the destination, then it can delivers packets on a much smaller number of hops. Of course the delivery delay is much longer, but the network capacity also increases by slowing non urgent packets. This is the time capacity paradox: by delaying packets, nodes mobility increases network capacity. This was hinted the first time by Grossglauser and Tse in 2002.

All of these results have been obtained assuming very strict assumptions of traffic density uniformity and ergodic motion. In this paper we will depart from the uniform models and assume that the traffic density varies with node location. We will provide results on how shortest path are affected by traffic density gradients. In particular we will show that in asymptotic conditions the routing paths obey to similar laws as in non linear optic. Therefore the scaling property of Gupta and Kumar is very general and can be interpreted by the inherent ability of wireless network to create capacity from the spatial density of nodes: this is the space capacity paradox of wireless networks.

We also generalise this result to the case where nodes are mobile and can hold the packet for some time before retransmitting it. Relaxing time constraint in the packet delivery and introducing mobility pattern depending on node position, we generalise the equation we obtained for a stationary network and prove that mobility can actually sharply increase the network capacity.

In conclusion we can say that space-time is in theory a source of capacity in wireless networks. The problem is to find the appropriate algorithms that allow to benefit from these properties. In this chapter we will show that a shortest path algorithm such as OLSR[6] can fullfill these conditions.

12.2. Gupta and Kumar Scaling Property

In their reference paper on the capacity of mobile ad hoc networks, Gupta and Kumar[2] showed that in presence of traffic density of λ bit per time unit per square unit area, the typical radius of correct reception decays in $O(\frac{1}{\sqrt{\lambda}})$. This result assumes an uniform density model and quantity λ is the density of traffic *including* the load generated by packet that are retransmitted on their way to their destination on multihop paths. As a direct consequence the average number of hops needed to connect two arbitrary points in a bounded domain is therefore $O(\sqrt{\lambda}L)$, where L is the diameter of the network domain, since the distance must be divided by the radio ranges. As pointed out by Gupta and Kumar, this property has a strong implication in the evaluation of the maximum capacity attainable by a random node when the node density increases. If C is the capacity generated by each node and N is the number of nodes in the network, Gupta and Kumar found that the maximum bandwidth attainable is $C_N = O(\frac{1}{\sqrt{N \log N}})$. The order of magnitude is obtained via the following milestones: the density of traffic generated per unit square are is $O(CN)$. Let $r(C,N)$ be the typical radio range, thus the number of retransmissions needed to route a packet from its source to its destination is $O(\frac{1}{r(C,N)})$. The latter estimate, in turn, yields a traffic density (including retransmissions) of $O(\frac{CN}{r(C,N)})$. Therefore $r(C,N) = O(\sqrt{\frac{r(C,N)}{CN}})$, namely $r(C,N) = O((CN)^{-1})$. The average number of neighbour per node is $O(\pi r(C,N)^2 N)$; it should be larger than $\log N$ in order to guarantee connectivity, which leads to the estimate $C = O((N \log N)^{-1/2})$.

Kumar and Gupta estimates were originally derived from information theory considerations and are not related to any particular network implementation. If we assume a specific implementation and propagation model, then there will be a quantity β such that the typical radius of correct reception of a packet is equal to $\frac{\beta}{\sqrt{\lambda}}$. By typical radius we mean the radius below which probability of correct reception of a packet is above a given threshold. The quantity β will depend on many parameters such as the probability threshold, the attenuation coefficient of wave propagation and the minimum signal-over-noise ratio required for correct reception. Notice that the typical disk of correct reception contains in average a finite number of transmitters per time unit, since the area is proportional to $\frac{1}{\lambda}$.

If we consider a network dispatched in a domain of dimension D then the estimate of the radius will be $\frac{\beta}{\lambda^{1/D}}$. In the sequel we will look at 2D domains generalising occasionnaly the results on other dimensions.

When the density λ increases in a fixed domain, then the minimum number of hops connecting two points A, B tends to be equivalent to $d(A,B)\frac{\sqrt{\lambda}}{\beta}$ where $d(A,B)$ denotes the euclidian distance between mobile node A and mobile node B. Meanwhile, the increase of the number of relays naturally increases the traffic density. If ν is the actual traffic density generation per unit area, *i.e.* the traffic locally generated on mobile nodes, not the traffic relayed by the mobile nodes, then the average density traffic will satisfy the identity: $\lambda = \nu \bar{d}\frac{\sqrt{\lambda}}{\beta}$ where $\bar{d}$ is the average euclidian distance between two end points in a connection.

This previous identity assumes that the pattern of path between points covers the domain in an uniform manner so that the traffic density, generated and relayed, is constant on the whole domain. In this case the path that connect two points with the minimum number of hops is very close to the straight line. But the question arises about the shape of the shortest path when the traffic density is not uniform. We will show that when the density increases while keeping proportional to a given continuous function, then the propagation paths tend to conform to continuous line, that we call propagation lines. Under these assumptions we will provide the general equations that the propagation lines must satisfy. We will show that variable traffic densities affect shortest path the same way as variable optical indices affect light path in a physical medium.

12.3. Massively Dense Networks

We now consider massively dense networks on a 2D domain. We denote by $\lambda(x,y)$ the traffic density at the point of coordinate (x,y) on the domain. We suppose that function $\lambda(x,y)$ is continuous in (x,y), or at least Lebesgue integrable. When $\lambda(x,y)$ are uniformly large, the results of Gupta and Kumar together with the result of the previous section state that the radio ranges tend to be "microscopic" and routes can be considered as continuous lines between nodes. Packets travelling on a route $\mathcal{C}$ passing on the point of coordinate (x,y) will experience hops of length $\frac{\beta}{\sqrt{\lambda(x,y)}}$ passing in the vicinity of point (x,y). Let $n(x,y) = \frac{\sqrt{\lambda(x,y)}}{\beta}$. The number of hops that a packet will experience on route $\mathcal{C}$ is something close to $\int_{\mathcal{C}} n(x(s),y(s))ds$ where s is a curvilign absciss on route $\mathcal{C}$.

In the sequel we are looking for route with the shortest hop number. Searching the path that minimise the hop number between two points A and B is therefore equivalent for looking for the path light between A and B in a medium with non-uniform optical index $\lambda(x,y)$. There is a known result about the optimal path that minimise a path integral $\int_{\mathcal{C}} nds$.

Theorem 12.1: The optimal path satisfies on each of its point $\mathbf{z}(s) = (x(s), y(s))$ such that s is acurvilign absciss ($ds = \sqrt{(dx)^2 + (dy)^2} = |d\mathbf{z}|$):

$$\frac{d}{ds}(n(\mathbf{z}(s))\frac{d\mathbf{z}(s)}{ds}) = \nabla n(\mathbf{z}(s)) \qquad (1)$$

where ∇ is symbol of gradient vector.

The proof is classical. If we consider a small perturbation $\mathcal{C}^*$ of optimal path $\mathcal{C}$ where $\mathbf{z}^*(s) = \mathbf{z}(s) + \delta \mathbf{z}(s)$ we should have $\int_{\mathcal{C}^*} nds^* - \int_{\mathcal{C}} nds = \delta[\int_{\mathcal{C}} nds] = 0$. We have $\delta[\int_{\mathcal{C}} nds] = \int_{\mathcal{C}} \delta[nds]$. Since $\delta[n] = \nabla n.\delta \mathbf{z}(s)$ and $\delta[ds] = \frac{d\mathbf{z}}{ds}.d\delta \mathbf{z}$ we get

$$\delta[\int_{\mathcal{C}} nds] = \int_{\mathcal{C}} \nabla n.\delta \mathbf{z}(s)ds + \int_{\mathcal{C}} n\frac{d\mathbf{z}}{ds}.\frac{d\delta \mathbf{z}}{ds}ds$$

Integrating by part the second right hand side integral of the above and assuming that both $\mathcal{C}$ and $\mathcal{C}^*$ share the same end points (*i.e.* $\delta\mathbf{z}(s) = 0$ at both ends), we get:

$$\delta[\int_\mathcal{C} n ds] = \int_\mathcal{C} (\nabla n - \frac{d}{ds}(n\frac{d\mathbf{z}}{ds})).\delta\mathbf{z}(s)ds.$$

Since $\delta\mathbf{z}(s)$ can be arbitrary, and that in all case $\delta[\int_\mathcal{C} n ds] = 0$, then $\nabla n - \frac{d}{ds}(n\frac{d\mathbf{z}}{ds}) = 0$ on the optimal path.

Therefore finding the optimal path is just an application of geometric optics. Notice that when $\nabla n = 0$ (uniform traffic density) propagation lines are straight lines (no curvature).

However we face a major problem in the fact that the distribution of path is actually impacting traffic density. This lead to an egg-and-chicken problem which may not be that easy to solve. We call $\Phi(x,y)$ the flow density of information transiting in the vicinity of point (x,y). Quantity $\Phi(x,y)$ is expressed in bit per meter, since it expresses the flow of packet crossing a virtual unit of segment of length of 1 meter centered on point (x,y). This flow impact the traffic density by the fact that each packet must be relayed every $\beta/\sqrt{\lambda(x,y)}$ meter in the vicinity of point (x,y). Therefore locally:

$$\lambda(\mathbf{z}) = \Phi(\mathbf{z})\frac{\sqrt{\lambda(\mathbf{z})}}{\beta} \qquad (2)$$

In other words $\lambda(\mathbf{z}) = (\frac{\Phi(\mathbf{z})}{\beta})^2$ and

$$n(\mathbf{z}) = \frac{\Phi(\mathbf{z})}{\beta^2} \qquad (3)$$

When considering domain of dimension D we have $\lambda = \Phi\frac{\lambda^{1/D}}{\beta}$ and $n = (\frac{\Phi}{\beta^D})^{\frac{1}{D-1}}$. Notice that the equations are singular when $D=1$.

As an example we can assume a planar domain massively and uniformly filled with mobile nodes and gateway nodes. We denote by μ_G the spatial density of gateways. We assume that the mobile nodes are much more dense than the gateways. We denote by ν the traffic density generated in any point. ν is expressed in bits per square meters per slot. The flow density Φ is constant in the domain and is equal to $\nu\bar{d}$. We suppose that mobile nodes sends and receives flows from their closest gateway. Therefore $\bar{d} = \int_0^\infty \exp(-\pi\mu_G r^2) dr = \frac{1}{2\sqrt{\mu_G}}$:

$$\Phi = \frac{\nu}{2\sqrt{\mu_G}} \qquad (4)$$

But in this case $\nabla n = 0$ and propagation lines are straight lines.

12.3.1. Tractable Case with Curved Propagation Lines

We exhibit two cases where the propagation lines are not straight lines. These cases are tractable because we will assume that we compute shortest path for additional light traffics that do not significantly impact the network traffic density.

The central gateway We assume a dense wireless network inside a disk. The main traffic comes from the boundary to the center G of the disk, where we assume there is a gateway to the wired backbone Internet. Assuming that the source of traffic are uniformly dispatched on the external circle and hinting that the propagation lines to the gateways are the straight lines, the flow density on a point **z** inside the disk is of the form

$$\Phi(\mathbf{z}) = \frac{\alpha}{|\mathbf{z}|}, \qquad (5)$$

for some α. Using Theorem 12.1 (see Ref. 5 for details), it comes that the shortest path between two points A and B are the branch of logarithmic spiral centered on G that connect A and B (see Fig. 12.2). When node A and B are at same distance to center G, then the shortest path is the fraction of arc that contains A and B. Consequently when A and B are symmetric with respect to point G, then surprisingly the shortest path starts orthogonally to the straight line connected the two points.

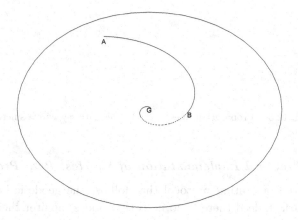

Fig. 12.2. Propagation lines in the central gateway wireless network.

The border gateways In this case we assume that the network contains multiple gateways to the backbone internet and all are located on the external circle. We assume that the gateways are uniformly dispatched on the circle. We also assume that the other nodes in the network are uniformly distributed in the disk and that most of their traffic are addressed to the external internet via the gateways. To this end every node selects the closest gateway, *i.e.* the gateway which is on the radius line that starts from center G and contains the node. Assuming the flows are straight lines to the gateways, the flow density satisfies:

$$\Phi(\mathbf{z}) = \alpha|\mathbf{z}| \ . \tag{6}$$

The equations of Theorem 12.1 can again be solved and it turns out that the propagation lines between two point A and B is

- an hyperbol branch if the angle between GA and GB is smaller than $\frac{\pi}{2}$ (see Fig. 12.3);
- the union of segment AG and GB, otherwise.

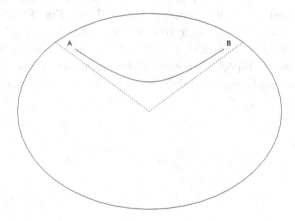

Fig. 12.3. Propagation lines in the border gateway wireless network.

12.3.2. *Practical Implementation of Shortest Path Protocol*

Implementing a routing protocol that follows the geodesic lines is not a difficult task. Indeed there is no need that nodes monitor their local traffic density n nor to advertise the gradient vector. In fact a shortest path algorithm, such as OLSR[6], will suffice. Of course one will need to limit the neighbourhood of the node to those whose hello success rate exceeds p_0. To this end we make use of Hysterisis Strategy in *advanced link sensing* option and set up HYST-THRESHOLD-HIGH parameter to value p_0 that provides the best success rate, hop distance compromise. Tuned that way OLSR will automatically provide the shortest path that fit the traffic density gradient curvature.

12.4. Introduction of Time Component

In the previous section we were assuming very strict timing constraints so that packets are forwarded like hot potatoes without any pause between retransmissions. Recently Grossglauser and Tse[4] showed that mobility increases the capacity of wireless ad hoc networks. This due to the fact when nodes are moving one just wait that nodes come closer instead of immediately starting relaying when nodes are far apart. In particular the increase in capacity can be dramatic (in $O(1)$ instead of $O(\lambda^{-1/2})$) if one considers ergodic mobility patterns. Unfortunately the delay for packet becomes unbounded when the density increases. The aim of this section is to quantify the gain in retransmissions number while we let the time constraint on delay delivery T vary.

Although basic sensors networks or other smart dust are not expected to be mobile, however one can imagine more sophisticated sensors produced by nanotechnology that can be mobile by themselves. The sensors can also travel because they are embedded in a mobile device or because the background medium is mobile (think about a sensor in a river stream) Interestingly intermediate nodes can choose to store packet while moving instead of immediately retransmitting it. As long as they move in the right direction this may considerably reduce the average number of retransmissions between source and destination. Of course the consequence is that packet delay delivery will considerably increase. This may be a solution for non urgent background traffic to take advantage of mobility and therefore have much less impact on global network load. Therefore we will model this very property by introducing space-time considerations in the framework presented in this paper.

Throughout this section we will assume that a node has a packet (or a sequence of packets) to transmit to a destination node with a time delivery constraint of T. In other words, each packet should arrive at its destination no later than after a delay T. It basically means that we add the time dimension to our 2D problem. A path now contains the time dimension and will connect a source space-time point $(\mathbf{z}_0, t_0)$ to a destination space-time point $(\mathbf{z}_1, t_1)$, given that $t_1 = t_0 + T$. When $T = 0$, and neglecting propagation delays and processing in relay nodes, we get to our previous analysis restricted to space components. Our aim is to show that with some mobility models, when T tends to infinity, the number of retransmissions needed to connect point $(\mathbf{z}_0, t_0)$ to point $(\mathbf{z}_1, t_1)$ tends to be negligible compared to the number of retransmissions needed to connect point $(\mathbf{z}_0, t_0)$ to point $(\mathbf{z}_1, t_0)$, (*i.e.* same spatial point but zero delay).

In order to set up notations and convention for this very general problem, we will first start with an unrealistic mobility model.

In this first example, we make the assumption that a node which has a packet to transmit or to relay can also travel with its packet at speed v. Therefore at any time the node that carries the packet has to make the decision of either transmitting it to the next hop or to travel with it during a certain duration. When the node chooses to hold the packet, we say that the packet is in hold state. We consider the optimal path $\mathcal{C}$ when $T = 0$ which connects $\mathbf{z}_0$ to $\mathbf{z}_1$, we assume that the space time path will be the path $\mathcal{C}$ plus a time component. In order to avoid too many notations we will still denote the space-time path by $\mathcal{C}$.

With these very hypotheses, the number of actual hops the packet will experience during its propagation on path $\mathcal{C}$ is equal to $\int_{\mathcal{C}} n |d\mathbf{z} - \mathbf{v} dt|$ where $\mathbf{v}$ is the vector speed at point $\mathbf{z}$. In the following we call $h = \frac{dt}{|d\mathbf{z}|}$ the average (local) packet holding time per distance unit, or we will denote by $\gamma = \frac{h}{v}$ the average fraction of distance traveled in hold state by a packet per distance unit.

In the sequel we consider a mobility model where the nodes are subject to random walks that are independent to data traffic conditions. In this model we assume that at any time node travels at a random speed toward a random direction and keep its speed and heading during a time duration τ. After time τ it randomly change speed and heading. This like a particle in motion in a gas. Quantity τ refers to the free space motion delay during which the particle moves in straight line. At the end of period τ the particle experiences a collision that changes its motion vector in a random way. Notice that τ can be made random as well (we may assume that it is exponentially distributed). We assume that the expectation of speed vector $E[\mathbf{v}] = 0$. We also assume that the speed vectors have isotropic direction and $c_2 \mathbf{I}$ is the covariance matrix, $\mathbf{I}$ being the identity matrix. We could accept some un-isotropic aspects so that covariance could depart from colinearity with identity matrix, but we will not do it for the sake of presentation.

Quantity $\sqrt{c_2 \tau}$ is the standard deviation of node location after one free space travel. We assume that this quantity is of the same order as of hop distance r (remaining that $r = \frac{1}{n}$). In other words the free space travel distance is of the same order as the hop distance. It is also instrumenetal in our proof that the speed is distributed on values that are not bounded by a finite number.

When a packet arrive in a mobile node, the router has to select whether it will transmit the packet to the next hop or keep it in hold state. We define a decision process which is based on the localisation of the next hop and the speed vector of the host node. If the node decides set the packet in hold state it will keep it as long its speed does not change. Therefore a hold state will last at least one τ period. The decision making automaton use a parameter x that is a positive real number and which depends on the delivery delay constraint T of the packet. Let θ be the angle made by the direction to the next hop. The node decide to immediately transmit its packet to the next hop iff the two following conditions hold:

(1) its speed is larger than parameter x;
(2) the speed direction angle is contained in interval $[-\theta - \frac{\pi}{1+x}, -\theta + \frac{\pi}{1+x}]$;

otherwise the packet stays in hold state. If the node keeps the packet in hold state it will keep it to its next motion vector change. At this moment it will proceed to a new packet state decision according to its new motion vector and to the localisation of the current potential next hop. If the node has been decided to be transmitted immediately then the receiver will also proceed to a state selection. A packet may be transmitted over several hops before returning back in hold state. Basically when $x = 0$, then the packet is always immediately retransmitted to its next hops as with $T = 0$; and when $x \to \infty$, then the packet is less likely retransmitted and stay longer in hold state. The probability that a node chooses to immediately transmit the packet is $p(x) = \frac{1}{1+x} P(|\mathbf{v}| > x)$ and let $v(x) = E[|\mathbf{v}|, |\mathbf{v}| > x] \frac{1}{\pi p(x)} \sin(\frac{\pi}{1+x})$, it is clear that $v(x) > \frac{x}{\pi} \sin(\frac{\pi}{1+x})$, since the computation is done on speed greater than x with direction uniformly distributed in $[-\theta - \frac{\pi}{1+x}, -\theta + \frac{\pi}{1+x}]$. The average motion vector when the packet is in hold state is colinear with the direction to the next hop and has modulus equal to $p(x)v(x)$. According to hypothesis we have $\lim_{x \to \infty} p(x) = 0$ and $\lim_{x \to \infty} v(x) = \infty$.

The average distance the packet will travel before a new decision has to be taken (either in hold state or in immediate retransmission) is $p(x)r + (1-p(x))p(x)v(x)\tau$ with variance $v_2(x)$. Notice that $\lim_{x\to\infty} v_2(x) = c_2\tau^2$. It comes that the average fraction of hold state travel per unit distance is

$$\gamma = \frac{(1-p(x))v(x)\tau}{r + (1-p(x))v(x)\tau} \qquad (7)$$

We have clearly $\lim_{x\to\infty} \gamma = 1$ since $v(x)$ tends to infinity, in this case we also have

$$T = \int_{\mathcal{C}} \frac{(1-p(x))v(x)\tau^2}{(r + (1-p(x))v(x)\tau)^2 p(x)} |d\mathbf{z}| \qquad (8)$$

which also tends to infinity since the product $p(x)v(x)$ tends to zero. However we cannot be sure that the packet actually reaches its destination. We know that the packet is *on average* on the path $\mathcal{C}$. In order to check how far from path it is actually we have to look at the variance of packet localisation. After each decision step, the packet travels in average a distance $p(x)r + (1-p(x))p(x)v(x)\tau$ with a variance $v_2(x)$. Therefore in order to travel a distance of one unit the packet will have go through an average number of decision steps equal to $\frac{1}{p(x)r+(1-p(x))p(x)v(x)\tau}$. Therefore the variance of its position is close to $\frac{v_2(x)}{p(x)r+(1-p(x))p(x)v(x)\tau}$. In order to be safe we have to prove that this variance is small, so that the packet does not evade too far from the path and that when time limit will be critical the number of hops it will have to travel in emergency to reach its destination won't be too long. Since $v_2(x) \sim c_2\tau^2$ which is of order of r^2 the variance is at most equal to $\frac{c_2\tau^2}{p(x)r}$ which is order $r/p(x)$ which is small. In other words the number of hops is (using identity $r = \frac{1}{n}$):

$$\int_{\mathcal{C}} \frac{n}{1+n(1-p(x))v(x)\tau n} |d\mathbf{z}| + O(|d\mathbf{z}|\sqrt{\frac{n}{p(x)}}) \qquad (9)$$

Similarily the delivery delay $T = \int_{\mathcal{C}} \frac{(1-p(x))v(x)\tau^2 n^2}{(1+(1-p(x))v(x)\tau n^2)^2 p(x)} |d\mathbf{z}|$. When the parameter $v(x)\tau$ is large compared to $1/n$ then the number of hops is equivalent to $\int_{\mathcal{C}} \frac{|d\mathbf{z}|}{v(x)\tau}$ which is in $O(\sqrt{\lambda}L)$ and $T \sim \int_{\mathcal{C}} \frac{|d\mathbf{z}|}{p(x)v(x)}$ which is in $O(\frac{L}{\sqrt{c_2}})$, where L is the physical diameter of the network. Notice that the optimal path may vary when T changes, for example if mobility model is uniform on the network domain then when T is large optimal path will be straight lines. In other word the curvature of optimal paths may also depend on the time component. Of course all the quantities x, $v(x)$ and $p(x)$ may also vary on the spatial domain leading to further optimisation.

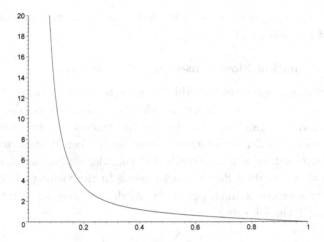

Fig. 12.4. Delay T versus the reduction factor of the number of retransmissions when $P(|\mathbf{v}| > x) = \frac{1}{1+x^2}$.

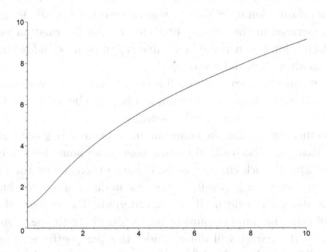

Fig. 12.5. Capacity gain factor versus Delay T for mobile nodes.

Figure 12.4 shows the gain in retransmission to the destination, and similarly the gain in capacity that can be achieved by tuning the factor x. When $x \to \infty$ we have $T \to \infty$, T being a fraction of $\frac{L}{E[|\mathbf{v}|]}$. For this plot we have taken a speed distribution in power law: $P(|\mathbf{v}| > x) = \frac{1}{1+x^2}$. In the next Fig. 12.5 we display the capacity gain factor with respect to delay T. Notice that the time delay unit coincides with the average time a node would take to carry its packet to its destination if it were driven straight instead of a random walk.

12.5. Information Flow Tensor and Perspectives

The previous examples are tractable because the path to the gateways were determined *a priori* by symmetry considerations and were not disturbed by the variation of parameter $n(\mathbf{z})$. In fact the problem is that propagation lines and the traffic densities mutually affect each other. We have to describe the information flow in more details. Let consider a point $\mathbf{z}$, let $\mathbf{u}_i$ be the sequence of information flow vectors passing in the vicinity of this point, each vector is expressed in bit per meter. We define the tensor of information flow vector $\mathbf{T}$ the following:

$$\mathbf{T}(\mathbf{z}) = \sum_i \frac{1}{|\mathbf{u}_i|} \mathbf{u}_i \otimes \mathbf{u}_i \qquad (10)$$

The definition is equivalent to the tensor stress-energy in physics. We have $\Phi = \text{tr}(\mathbf{T}) = \sum_i |\mathbf{u}_i|$. The divergence of $\mathbf{T}$ can be interpreted as local generation of information flow: $\nabla \cdot \mathbf{T} = \mathbf{v}$ where $\mathbf{v}$ is the sum of all information vectors generated in the vivinity of point ($\mathbf{z}$). An information vector $\mathbf{v}$ is when information is generated in volume $|\mathbf{v}|$ on point $\mathbf{z}$ and is transmitted toward a path with initial vector $\frac{\mathbf{v}}{|\mathbf{v}|}$.

It seems also that propagation lines don't change when the route optimisation criterium changes. For example if hop number is changed in packet total delay time, the route should basically remains the same. The reason for this conjecture is that the condition of traffic at any given point in the network location is basically the same *modulo* an homothetic factor $\lambda(\mathbf{z})$. The only aspect which changes is the distance travelled by the packet per hop, but the delay per hop will be the same in distribution. We have a similar point about bandwidth allocation criterium. The transmission at any point will take the same amount of bandwidth, only the per hop distance travelled by the packet will differ. Under this perspective simple shortest path algorithms such as OLSR[6] should be asymptotically close to optimal.

References

1. Jacquet, P., "Elément de théorie analytique de l'information, modélisation et évaluation de performances," *INRIA Research Report* (1998) RR-3505. http://www.inria.fr/rrrt/rr-3505.html
2. Gupta, P. and Kumar, P.R., "Capacity of wireless networks" *Technical report, University of Illinois, Urbana-Champaign*, (1999). http://citeseer.nj.nec.com/gupta99capacity.html

3. Baccelli, F., Blaszczyszyn, B., Mühlethaler, P., "A spatial reuse Aloha MAC protocol for Multihop wireless mobile networks" *INRIA Research Report* (2003) RR-4955. http://www.inria.fr/rrrt/rr-4955.html
4. Grossglauser, M., Tse, D., "Mobility increases the capacity of ad hoc wireless networks", 2001, 1360-1369, (INFOCOM, 2001).
5. Jacquet, P., "Geometry of information propagation in massively dense ad hoc networks" INRIA research repport RR-4992, 2003.
6. Clausen, T., Jacquet, P., "The Optimized Link State Routing protocol (OLSR)" IETF RFC 3626, 2003.
7. Adjih, C., Baccelli, E., Clausen, T., Jacquet, P., Rodolakis, R., "Fish Eye OLSR Scaling Properties" *Journal of Communications and Networks*, Special Issue on Mobile Ad Hoc Networks, (Dec. 2004).

CHAPTER 13

Stochastic Properties of Peer-to-Peer Communication Architecture in a Military Setting

Donald P. Gaver and Patricia A. Jacobs

Department of Operations Research, Naval Postgraduate School
1411 Cunningham Road, Monterey CA 93943-5214, USA
E-mail: dgaver@nps.edu, pajacobs@nps.edu

A time-critical military mission must be completed before a random deadline to be successful. The mission requires a number of Blue assets; the more assets that can be assembled before the deadline, the greater the possibility of mission success. Traditionally, military Command and Control (C2) has a hierarchical structure; information is centrally stored and decision makers are also centralised. The paradigm of Network-Centric Warfare (NCW) implies a more horizontal C2 structure. There is often little communication infrastructure on the battlefield. Peer-to-Peer (P2P) communication networks are attractive enablers of a horizontal C2 structure. A stochastic model is used to discuss the benefits and possible vulnerabilities of a P2P-enabled C2 structure for a time-critical mission. The P2P architecture can result in larger probabilities of mission success than centralised C2. However, its benefits are nullified if the time it takes to assemble the needed Blue assets becomes larger than that for the centralised C2.

13.1. Problem Formulation

We consider here the situation in which a single opponent Red Agent (RA) moves surreptitiously in a designated subregion, either on earth/terrain or on or below a maritime littoral surface. Its mission may be any of a variety of options. For instance, a TEL (Transportable Erectable Launcher) is a land based SCUD missile launcher that tries to elude detection by hiding under bridges or trees, or in caves, or even in partially empty buildings, such as barns. It becomes visible (vulnerable to attack) when it sets up to shoot at remote Blue assets, or when it is in motion across terrain (along a road) in its assigned subregion or sector. Otherwise it is invisible (hiding or concealed). This type of target is called a *time-sensitive target*[1]. Alternatively, Blue is protecting a littoral domain, and an overhead sensor is searching for a Red (hostile and possibly lethal) platform attempting to reach, and damage, Blue assets. Overhead sensors identify the Red amongst a large background of harmless White/Friendly vessels; once identified (possibly incorrectly) reinforcement Blue surface (or air) platforms assemble to divert or destroy the suspected threat.

Another interpretation is that "the Red" is actually a friendly (Blue) downed pilot in hostile territory, and is moving about to avoid capture by opponent-Reds. Blue rescuers/recovery teams try to find and recover the pilot before hostile capture occurs.

Each subregion (one of many that comprise a larger total region) is assumed for the present to be covered by two Blue Agent (BA) types: there are n_s Blue Surveillance Agents (BSAs) and n_b Blue Attack Agents (BAAs); it is possible that the BAs are equipped both to see and to shoot, for example, be Un-Manned Combat Aerial Vehicles (UCAVs), or armed Fixed Wing Manned Aircraft, or combinations thereof. The BSAs search the subregion "at random," and then signal, either through a central C2 authority, or directly, for example, by "chat"[2], to loitering BAAs when the RA is detected. Once signaled, the BAAs are given the detected RA's location and proceed thereto.

Assume the rate of arrival near to Red's location of any BAA after being signaled is Markovian with rate λ (which may depend on region or terrain), so the time a typical BAA requires finding and "attaching to" the visible RA, initiating the formation of an Engagement Pack (EP)[3], has an exponential distribution with mean $1/\lambda$ independently of the other BAAs. This formulation may be altered to reflect the particular behaviour of various communication-network-forming (Peer-to-Peer (P2P)) protocols[4-7]. When an EP has reached a to-be-determined size $\tilde{b}$ it (the entire EP) attacks/fires upon the RA if the RA remains visible. If the RA senses an EP, or if it finishes its assigned mission before the EP($\tilde{b}$) fully forms, it immediately hides; the BSA-BAA mission is a failure, and the Red-Hide, Blue-Seek process begins again. There are certain tradeoffs: if $n_s \gg n_b$ then a visible RA may be found quickly, but relatively small EPs may be optimal because larger assemblies could alert the visible Red to hide; this lowers the lethality of B on R (but deters hostile SCUD launches). It is clear that if $\tilde{b} \leq n_b$ is too small, low Blue-on-Red attack lethality may occur, while if $\tilde{b}$ is set too large the chance increases that the visible RA (potential target) senses the threat, or hastily completes its mission, and disappears. There can clearly be an optimal EP size, and the latter may realistically evolve over time as learning by both sides occurs. Also, the latter may depend on physical/environmental conditions, such as ducting[8].

13.2. A Renewal Model for Blue vs Red in a Subregion

Let there be n_s BSAs, and suppose each independently searches for a visible RA over the distinguished subregion, and let a typical random visible period for the RA be of independent random duration $\mathbf{V}$ with distribution $F_\mathbf{V}$ with $F_\mathbf{V}(0) = 0$; (it is realistic that Red may alter the latter over time and experience with near-misses by Blue, but we do not address this gaming aspect of the problem at this stage). If the RA becomes visible (at $t = 0$), then some BSA detects it at time $\mathbf{D} = x$ if $\mathbf{D}(= x) < \mathbf{V}$; let the Poissonian search rate of the BSA force be $\bar{\xi} = n_s \xi$; independence then implies that the probability the RA is detected before it hides is

$$p_d(BR) = \int_0^\infty [1 - F_\mathbf{V}(x)]\bar{\xi} e^{-\bar{\xi} x}\, dx = 1 - \hat{F}_\mathbf{V}(\bar{\xi}) \qquad (1)$$

where

$$\hat{F}_\mathbf{V}(\bar{\xi}) = \int_0^\infty e^{-\bar{\xi} x} F_\mathbf{V}(dx) \equiv E[e^{-\bar{\xi} \mathbf{V}}]; \qquad (2)$$

coincidentally the Laplace-Stieltjes transform for the distribution function of $\mathbf{V}$.

Let $\mathbf{H}$ with distribution function $F_\mathbf{H}$ having Laplace-Stieltjes transform $\hat{H}(s)$ denote an arbitrary hiding time; and let $\{\mathbf{H}(n)\}$ and $\{\mathbf{V}(n)\}$ be mutually independent sequences of independent identically distributed random variables. It is easily possible to allow their distributions to be conditional on, say, environmental factors (ducting, sea state, terrain, etc.) and on the actual types of RAs and BAs in play.

13.2.1. The Visibility Detection Process

The temporal history of a RA can be (initially) modelled and analysed, for example, by probabilistic mathematics or Monte Carlo simulation, as an alternating renewal process[9]. Suppose the RA initially begins hiding. Then assume it is in hiding during the random intervals

$$[0, \mathbf{H}(1)], [\mathbf{H}(1) + \mathbf{V}(1), \mathbf{H}(1) + \mathbf{V}(1) + \mathbf{H}(2)], \ldots.$$

It is visible on the complementary intervals

$$[\mathbf{H}(1), \mathbf{H}(1) + \mathbf{V}(1)],$$

$$[\mathbf{H}(1) + \mathbf{V}(1) + \mathbf{H}(2), \mathbf{H}(1) + \mathbf{V}(1) + \mathbf{H}(2) + \mathbf{V}(2)], \ldots.$$

To model detection of the RA requires the BSA surveillance capability, which is represented as occurring at rate $\bar{\xi} = n_s \xi$ over the subregion.

13.2.2. The Model

Take this convenient if simplified approach to BSA detection. If an element of BSA force arrives when the RA is hiding, no detection of the RA is possible, whereas if the hidden RA becomes visible (for time $\mathbf{V}$) then assume that its detection occurs if a BSA unit arrives near the RA location at time $\mathbf{D}$ after the RA emerges and before it hides again. The probability of this event is given by (1) above (assume that the RA does not detect the BSA and immediately hide; such capability could induce a more surreptitious BSA concept of operations (CONOPS)).

Let $\mathbf{T}_D$ denote the random first BSA detection time of a visible RA, given that RA has initially become visible at time 0. Then

$$\mathbf{T}_D = \begin{cases} \mathbf{D} & \text{if } \mathbf{D} < \mathbf{V}(1), \\ \mathbf{V}(1) + \mathbf{H}(1) + \mathbf{T}'_D & \text{if } \mathbf{D} > \mathbf{V}(1) \end{cases} \qquad (3)$$

where $\mathbf{T}'_D$ is an independent replica of $\mathbf{T}_D$. It is straightforward to calculate the Laplace-Stieltjes transform $E[e^{-s\mathbf{T}_D}]$:

$$\psi_{\mathbf{T}_D}(s) = \frac{\bar{\xi}[1 - \hat{F}(s + \bar{\xi})]}{(s + \bar{\xi})[1 - \hat{F}(s + \bar{\xi})\hat{H}(s)]} \qquad (4)$$

where $\bar{\xi} = \xi n_s$. Setting $s = 0$ shows that $\mathbf{T}_D$ terminates in finite time with probability one. The Laplace transform of $P\{\mathbf{T}_D > t\}$ is

$$\frac{1 - \psi_{\mathbf{T}_D}(s)}{s} = \frac{s[1 - \hat{F}_{\mathbf{V}}(s + \bar{\xi})\hat{H}(s)] + \bar{\xi}\hat{F}_{\mathbf{V}}(s + \bar{\xi})[1 - \hat{H}(s)]}{s(s + \bar{\xi})[1 - \hat{F}_{\mathbf{V}}(s + \bar{\xi})\hat{H}(s)]}$$

$$= \frac{1}{s + \bar{\xi}} + \frac{\bar{\xi}}{(s + \bar{\xi})} \frac{\hat{F}_{\mathbf{V}}(s + \bar{\xi})}{[1 - \hat{F}_{\mathbf{V}}(s + \bar{\xi})\hat{H}(s)]} \frac{1 - \hat{H}(s)}{s}. \qquad (5)$$

Let $s \to 0$ to see that

$$E[\mathbf{T}_D] = \frac{1}{\bar{\xi}} + \frac{\hat{F}_{\mathbf{V}}(\bar{\xi})}{(1 - \hat{F}_{\mathbf{V}}(\bar{\xi}))} E[\mathbf{H}]. \qquad (6)$$

On the set $\mathbf{D} < \mathbf{V}$ let $\mathbf{V}_0 = \mathbf{V} - \mathbf{D}$, the remaining time the RA is visible. The Laplace transform of the remaining time the RA is visible on the set the RA is detected before it hides is

$$E[e^{-s\mathbf{V}_0}; \mathbf{D} < \mathbf{V}] = \int_0^\infty \bar{\xi} e^{-\bar{\xi} y} \, dy \int_y^\infty e^{-s(z-y)} F_\mathbf{V}(dz)$$

$$= \frac{\bar{\xi}}{\bar{\xi} - s} \hat{F}_\mathbf{V}(s) - \frac{\bar{\xi}}{\bar{\xi} - s} \hat{F}_\mathbf{V}(\bar{\xi}); \qquad (7)$$

$$\phi(s) = E[e^{-s\mathbf{V}_0} | \mathbf{D} < \mathbf{V}] = \frac{\bar{\xi}}{\bar{\xi} - s} \frac{[\hat{F}_\mathbf{V}(s) - \hat{F}_\mathbf{V}(\bar{\xi})]}{1 - \hat{F}_\mathbf{V}(\bar{\xi})}. \qquad (8)$$

This development assumes that just one detection by a BSA (the first such) is sufficient to begin to enroll an Engagement Pack (capable of lethal action, or possibly pilot recovery). Otherwise, Peer-to-Peer communication could be used to develop a corroborative group of observers[10–12].

13.2.3. The Probability an EP of Size $\tilde{b}$ Attaches to the Detected RA Before it Hides

Assume once the RA is detected, BAAs attach to it after independent exponential times each having mean $1/\lambda$ while the RA is visible. Note that more rapid EP assembly may be achieved by using a more sophisticated and situationally adapted P2P discipline. Many alternatives are possible and adaptable to changed circumstances; these are under investigation.

The conditional probability an EP of size $\tilde{b}$ forms before the RA hides given $\mathbf{V}_0$ is

$$P\{\mathbf{T}_{\tilde{b}} < \mathbf{V}_0 | \mathbf{V}_0, \mathbf{D} < \mathbf{V}\} = \sum_{k=\tilde{b}}^{\bar{b}} \binom{\bar{b}}{k} [1 - e^{-\lambda \mathbf{V}_0}]^k [e^{-\lambda \mathbf{V}_0}]^{\bar{b}-k}$$

$$= \sum_{k=\tilde{b}}^{\bar{b}} \binom{\bar{b}}{k} \sum_{j=0}^{k} \binom{k}{j} (-1)^j e^{-\lambda \mathbf{V}_0 [j + \bar{b} - k]} \qquad (9)$$

where $\mathbf{T}_{\tilde{b}}$ is the time to form an EP of size $\tilde{b}$ and $\bar{b}$ is the number of BAAs in the subregion for $\tilde{b} \leq \bar{b}$.

13.2.4. The Probability the Detected RA is Killed Before it Hides

Given $\bar{b}$, the number of BAAs in a region, we calculate the probability of Blue's successful kill of a discovered RA. Mathematically, this amounts to

$$P_K(\tilde{b},\bar{b}) = E_{\mathbf{V}_0|(\mathbf{D}<\mathbf{V})}\left[P\{\mathbf{T}_{\tilde{b}} \leq \mathbf{V}_0|\mathbf{V}_0\}(\bar{h})^{\tilde{b}}(1-(1-p_K)^{\tilde{b}})\right] \quad (10)$$

where $\mathbf{T}_{\tilde{b}}$ is the time of first passage to $\tilde{b}$ present in an EP, $\tilde{b} \leq \bar{b}$, and $\bar{h}$ is the probability that an individual BAA's arrival does not cause the RA to hide. The term $1-(1-p_K)^{\tilde{b}}$ denotes the probability that if all $\tilde{b}$ BAAs attached to the RA shoot, at least one hits/kills the RA; this assumes independence and simultaneous independent Blue firing success probabilities, all of which can be modified for more realism. The conditional probability the RA is killed before it hides given the RA is detected during a visible period is thus

$$P_K(\tilde{b},\bar{b}) = (\bar{h})^{\tilde{b}}\left(1-(1-p_K)^{\tilde{b}}\right)\sum_{k=\tilde{b}}^{\bar{b}}\binom{\bar{b}}{k}\sum_{j=0}^{k}\binom{k}{j}(-1)^j\phi(\lambda[j+\bar{b}-k]) \quad (11)$$

where $\phi(s)$ is defined in (8).

13.2.5. Model with Additional C2 Time

In addition to the time to assemble an EP there can be an additional planning/command time, $\mathbf{S}$, which we assume has an exponential distribution with mean $1/\alpha$, independent of the time to assemble the EP. This assumption can be justified on the basis of heavy-traffic queuing theory[13]. This is the traditional C2/I or C4ISR Central Command Time/latency; it may conservatively ensure against errors, but adds a latency penalty to Blue response. A visible RA is engaged if the time to assemble the EP plus the planning/command time is less than the remaining time the RA is visible. See detailed modelling of a more complex version of this problem by Brickner[1].

$$P\{S + T_{\tilde{b}} < V_0 | V_0, D < V\}$$

$$= \sum_{k=\tilde{b}}^{\bar{b}} \binom{\bar{b}}{k} \int_0^{V_0} \alpha e^{-\alpha s}[1 - e^{-\lambda(V_0-s)}]^k [e^{-\lambda(V_0-s)}]^{\bar{b}-k} ds$$

$$= \sum_{k=\tilde{b}}^{\bar{b}} \binom{\bar{b}}{k} \sum_{j=0}^{k} \binom{k}{j} (-1)^j \int_0^{V_0} \alpha e^{-\alpha s} e^{-\lambda[j+\bar{b}-k](V_0-s)} ds$$

$$= \sum_{k=\tilde{b}}^{\bar{b}} \binom{\bar{b}}{k} \sum_{j=0}^{k} \binom{k}{j} (-1)^j \frac{\alpha}{\alpha - \lambda(\bar{b}+j-k)} \left[e^{-\lambda(\bar{b}+j-k)V_0} - e^{-\alpha V_0} \right].$$
(12)

In this case, the conditional probability the RA is killed before it hides, given it is detected, is

$$P_K(\tilde{b}, \bar{b}; a) = (\bar{h})^{\tilde{b}} \left(1 - (1 - p_K)^{\tilde{b}}\right)$$

$$\times \sum_{k=\tilde{b}}^{\bar{b}} \binom{\bar{b}}{k} \sum_{j=0}^{k} \binom{k}{j} (-1)^j \frac{\alpha[\phi(\lambda[j+\bar{b}-k]) - \phi(\alpha)]}{\alpha - \lambda(\bar{b}+j-k)}.$$
(13)

13.2.6. Numerical Illustration

In this example the time to detect a visible RA is given an exponential distribution with mean 4 minutes. The RA is visible for a random time having a gamma distribution with mean 15 minutes and shape parameter 50. The probability the target hides when a new BAA joins the EP is 0. Table 13.1 displays the probabilities a visible RA is killed when it is engaged by various sizes of engagement packs. The time for a BAA to join an engagement pack has an exponential distribution independent of the other BAAs.

Figure 13.1 displays the probability of killing the RA during its visible period for 3 different mean additional C2 times; the mean times are 0, 1.5 minutes, and 3 minutes. In all cases the mean time until a BAA joins an engagement pack is 3 minutes. For each EP size the number of BAAs in the region equals the needed size of the EP. The probability of killing the detected RA during its visible period is maximised for an engagement pack of size 3 for all cases. The probability the RA will hide before a 4th BA joins an EP of size 3 nullifies the benefit of the increased probability of kill that an EP of size 4 has. Increasing the mean C2 time decreases the probability of killing the RA during its visible period for all engagement pack sizes.

Table 13.1. Conditional probability an engaged visible target is killed given the size of the engagement pack.

Size of EP	Cond. prob. of kill
1	0.5
2	0.6
3	0.7
4	0.8
5	0.9

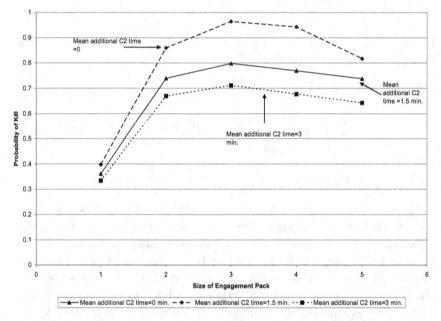

Fig. 13.1. Probability of killing a detected RA during its remaining visible period versus the size of the engagement pack. The length of the visible period has a gamma distribution with mean 15 min and shape 50. The time to detect a visible target has an exponential distribution with mean 4 min. The time for a BAA to join an engagement pack has an exponential distribution with mean 3 min. Note that the optimal EP size is about the same for all cases, but the actual kill probabilities are highly time-sensitive.

Figure 13.2 displays the probability the detected RA is killed during its remaining visible period for three cases with different mean times for a BAA to join an engagement pack: 3 minutes, 4.5 minutes, and 6 minutes;

the additional C2 time is equal to 0 for all three cases. For each EP size, the number of BAAs in the region is equal to the EP size. Note that the mean time to form an EP of size one for each case is the same as that in Fig. 13.1. The probability of killing the detected RA during its remaining visible time is maximised for an engagement pack of size 3. Increasing the mean time for a BA to join an EP decreases the probability of kill. The decrease becomes larger the larger the size of the engagement pack. Comparison with Fig. 13.1 reveals that for an engagement pack of size 5, increasing the mean time for a BAA to join an engagement pack results in a smaller probability of kill than increasing the mean additional C2 time for the cases considered.

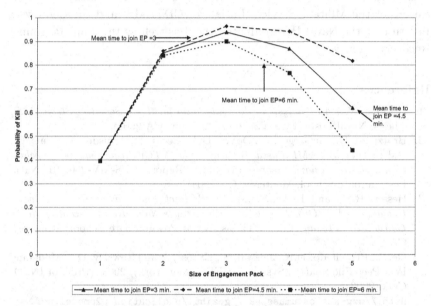

Fig. 13.2. Probability of killing the detected RA during its remaining visible period versus the size of the engagement pack. The length of the visible period has a gamma distribution with mean 15 min. and shape 50. The time to detect a visible target has an exponential distribution with mean 4 min. There is no additional C2 time.

13.3. Conclusions

A stochastic model for the effect of a P2P-enabled C2 architecture on the effectiveness of time-critical targeting has been presented. The numerical example suggests that P2P can result in increased targeting effectiveness over a centralised C2 architecture. However, if P2P results in an increased time to assemble an engagement pack to engage the target over that using a centralised C2 architecture or if there are too many errors or too much jamming or interference, then its benefits can be nullified. The model is optimistic in that it assumes that all BAAs are available to join an engagement pack when a target is detected and Red is not attempting to disrupt the P2P communication. The model can be modified to represent the consequences of different P2P communication networks.

Acknowledgments

This work was supported by the Center for Defense Technology and Education for the Military Services Initiative (CDTEMS) and the Cebrowski Institute at the Naval Postgraduate School as well as the Joint Personnel Recovery Agency.

References

1. Brickner, W., *An Analysis of the Kill Chain for Time Sensitive Strike* (M.S. Thesis, Naval Postgraduate School, Monterey, CA 2005).
2. Brutzman, D., McGregor, D., DeVos, D.A., Lee, C.S., Armsden, S., Blais, C., and Hittner, B., *XML-Based Tactical Chat (XTC): Requirements, Capabilities, and Preliminary Progress* (Technical Report, NPS-MV-2004-01, Naval Postgraduate School, Monterey, CA 2004).
3. Hesser, R.W. and Rieken, D.M., *FORCEnet Engagement Packs: "Operationalizing" FORCEnet to Deliver Tomorrow's Naval Network-centric Reach Capabilities...Today* (M. S. Thesis, Naval Postgraduate School, Monterey, CA 2003).
4. Ge, Z., Figueiredo, D.R., Jaiswal, S., Kurose, J. and Towsley, D., "Modelling Peer-Peer File Sharing Systems" (IEEE, New York), 2003, (Proc. of INFOCOM 2003).
http://www-net.cs.umass.edu/~gezihui/publication/p2pmodel.pdf.
5. Qiu D. and Srikant R., "Modelling and Performance Analysis of BitTorrent-Like Peer-to-Peer Networks" (ACM, New York), 2004, 367-368, (Proc. ACM SIGCOMM 2004).
http://comm.csl.uiuc.edu/~srikant/Papers/sigcomm04.pdf.
6. Manoski, S.,"Eliminating the Middleman: Peer-to-Peer Technology for Command and Control" *MITRE-The-Edge*, **6**(2) (The MITRE Corporation, Bedford, MA), 2002, 8-9.
http://www.mitre.org/news/the_edge/summer_02/manoski.html.
7. Gelenbe, E., Lent, R., Monturo, A., and Xu, Z., "Cognitive Packet Networks: QoS and Performance", (IEEE, New York), 2002, 3-12, (Proceedings of the IEEE MASCOTS 2002 Conference).
http://www.cs.ucf.edu/~erol/Mascots02.pdf.

8. Frederickson, P.A., and Davidson, K.L.,*An Operational Bulk Evaporation Duct Model* (Meteorology Dept., Naval Postgraduate School, Monterey, CA 2003).
9. Feller W., *An Introduction to Probability Theory and its Applications*, Vol. II (Wiley), 1966.
10. Anwar, Z., Yurcik, W., and Campbell, R.H.,"A Survey and Comparison of Peer-to-Peer Group Communication Systems Suitable for Network-Centric Warfare" in "Visual Information Processing XIV" Rahman, Z., Schowengerdt, R.A., and Reichenbach, S.E. ed (SPIE, Bellingham, WA) 2005, 33-44, (Proc. of the SPIE, Vol. 5820).
http://www.ncassr.org/projects/multicast/papers/spie05-p2p.pdf.
11. Gelenbe, E., Gellman M., and Loukas, G., "An Automatic Approach to Denial of Service Defence", (IEEE, New York), 2005, 537-541,(Proceedings of the IEEE International Symposium on a World of Wireless, Mobile and Multimedia Networks).
http://san.ee.ic.ac.uk/publications/dos_acc05.pdf.
12. Gelenbe, E. and Liu, P.,"QoS and Routing in the Cognitive Packet Network", (IEEE, New York), 2005, 517-521, (Proceedings of the IEEE International Symposium on a World of Wireless, Mobile and Multimedia Networks 2005).
http://san.ee.ic.ac.uk/publications/sp_acc05.pdf.
13. Whitt, W., *Stochastic-Process Limits* (Springer-Verlag, New York), 2002.

CHAPTER 14

Quantifying the Quality of Audio and Video Transmissions over the Internet: The PSQA Approach

Gerardo Rubino

INRIA / IRISA,
Campus de Beaulieu
35042 Rennes Cedex, France
E-mail: rubino@irisa.fr

Consider the problem of delivering multimedia streams on the Internet. In order to decide if the network is working correctly, one must be able to evaluate this delivered quality as perceived by the end user. This can be done by performing subjective tests on samples of the received flows, but this is expensive and not automatic. Being able to perform this task in an automatic way and efficiently, such that it can be done in real time, allows multiple applications, for instance in network control. Such a goal is achieved by our PSQA metric: Pseudo-Subjective Quality Assessment. It consists of training a Random Neural Network (RNN) to behave as a human observer and to deliver a numerical evaluation of quality, which must be, by construction, close to the average value a set of real human observers would give to the received streams. This chapter reviews our previous work on PSQA, and provides some new elements about this technology. The use of RNN is justified by its good performance as a statistical learning tool; moreover, the model has nice mathematical properties allowing using it in many applications and also obtaining interesting results when PSQA is coupled with standard modelling techniques.

14.1. Introduction

Consider an audio stream sent through an IP network, or a video one, or consider an interactive voice communication over the Internet. When quality varies, which is today very often the case (think of a wireless segment, for instance), it is useful to be able to quantify this quality evolution with time, in order to understand how the global communication system works, why the performance is as observed, how it can be improved, how the system can be controlled to optimise quality, etc. Observe that we are here interested in quantifying the quality *as perceived by humans users*. This task is so useful when analysing a networking application dealing with these kinds of flows that it has been standardised (some examples are for audio flows[1] and for video ones[2]). The corresponding area is called *subjective testing*. In a nutshell, it consists of taking a panel of human observers (say, around 10 to 20 persons sampled from the global population of observers, or a few experts, say 3 or 4, the choice between random observers or experts depending on the goals), and making them evaluate numerically the quality as they perceive it, by comparing an important number of sequences. The main problem with this approach is that it is expensive and, by definition, it is not an automatic process (and *a fortiori* not a real-time one).

If we try to perform this quality quantyfying process automatically, a first idea is to use *objective tests*. This basically refers to using techniques coming from coding, where the original sequence σ and the received one σ' are compared by computing some distance between them. A typical example of this is the PSNR metric (Peak Signal to Noise Ratio). It is well known that this approach does not correlate well with subjective ones, that is, with values coming from real human observers[a].

In a series of papers[3–5] we developed a new methodology called PSQA (see next section). It allows to reach the objective of being able to measure quality *as perceived by the users*, automatically and accurately. This chapter describes the approach and some of its applications. Next section describes PSQA. Section 14.3 is a short presentation of Random Neural Networks, the mathematical tool used to implement the PSQA technology, with some new elements concerning sensitivity analysis. In Section 14.4 we explain how we couple PSQA with standard performance evaluation techniques, close to the work presented by Rubino and Varela[5]. Section 14.5 concludes the chapter.

14.2. The PSQA Technology

PSQA stands for Pseudo-Subjective Quality Assessment, and it is a technology proposed first in Mohamed and Rubino[3] (under the name QQA: Quantitative Quality Assessent), in order to perform this quantifying task

---

[a]What we say is that objective measures are not good to quantify quality when considering flows that have been perturbed by travelling through a network where they can suffer from losses, delays, etc. They are of course very useful when analysing *coding* effects.

automatically, in real-time if necessary, and accurately. This last condition means that a PSQA metric must give to a stream a value close to the value an average human observer would give to it (or, in other cases, a pessimistic expert, or an optimistic one, depending on the goals of the study), as coming out of a subjective testing experiment. To describe how PSQA works, let us consider the following example, which a simplified version of the one used in Mohamed and Rubino[3]. We want to analyse a video streaming application, sending its flows from a source to a receiver. We proceed as follows:

(1) We first identify a set of parameters which, beforehand, we *think* have an (important) impact on the perceived quality. Observe that this is an *a priori* assumption. These parameters are of two classes: those related to the codec, which we assimilate to the source sending the stream, and those associated with the network transporting the data.

In our example, let us consider that the sending process can work at different bit rates BR, given in bps (bits per second), and at different frame rates, FR, in fps (frames per second). Assume that (i) the frames have all the same size in bits, (ii) there is no protection against losses (such as FEC – Forward Error Correction), (iii) the player at the receiver side can play a sequence with missing frames.

Concerning the network, we assume that the main problem is loosing packets since the receiver has a large buffer to absorb the possible variations in delay (jitter). For the loss process, given its importance we decide to characterise it by two parameters, the packet loss rate LR and the mean loss burst size MLBS, the latter giving an indication about how the losses are distributed in the flow.

(2) For each selected parameter we choose a range, corresponding to the system to be analysed and the goals; we also choose a discrete set of values, which make the rest of the process simpler. For instance, in Mohamed and Rubino[3] it is considered that LR $\in \{0\,\%, 1\,\%, 2\,\%, 3\,\%, 5\,\%, 10\,\%\}$. Each combination of values for these parameters is called a *configuration* of the system. The total number of configurations can be high (in our previous applications of PSQA we always had spaces with thousands of points).

(3) We perform a selection of M configurations among the set of all possible ones. Typically, for 4 to 6 parameters we used M around 100 to 150. This selection process is made by a merge of two procedures: (i) different

configurations are built by randomly choosing the parameters' values in the given space and (ii) several configurations are choosen by *covering* in some way the extremities of the parameters' ranges[3].

We randomly separate the set of M selected configurations into two subsets $\mathcal{C} = \{\gamma_1, \cdots, \gamma_K\}$ and $\mathcal{C}' = \{\gamma'_1, \cdots, \gamma'_{K'}\}$. We thus have $K + K' = M$ (for instance, in our applications, if $M \approx 100$ we will take, say, $K \approx 80$, or 85, and $K' \approx 20$, or 15. Set $\mathcal{C}$ will be used to *learn* and set $\mathcal{C}'$ to *validate* (see step (7)).

(4) We build a platform allowing (i) to send a video sequence through an IP connection, (ii) to control the set of parameters chosen in the first step; in the example, we must be able to choose *simultaneously* any values of the four parameters BR, FR, LR, MLBS.

(5) We choose a sequence σ representative of the population of sequences to be considered[b]; according to the norms used in video subjective testing, σ must be about 10 seconds length. We send σ exactly M times from sender to receiver, using the platform, and each time using the values in each of the selected configurations (sets $\mathcal{C}$ and $\mathcal{C}'$). We obtain two sets of *distorted copies* of σ: the copy corresponding to configuration γ_i, $1 \leq i \leq K$ (resp. γ'_j, $1 \leq j \leq K'$) is denoted by σ_i (resp. by σ'_j).

(6) We perform a subjective testing experiment using the M sequences $\sigma_1, \cdots, \sigma_K, \sigma'_1, \cdots, \sigma'_{K'}$. This consists of selecting a panel of human observers and asking them to numerically evaluate the quality of the M sequences *as they perceive them*. For this purpose an appropriate norm must be followed. In our video experiments, we used ITU-R Recommendation BT.500-10[2].

Each sequence will thus receive a value usually called MOS (Mean Opinion Score). We denote the MOS of sequence σ_i (resp. of σ'_j) by Q_i (resp. by Q'_j). In more detail, assume there are R observers (typically, $R \approx 20$ if the observers are random elements of the population, or $R \approx 4$ if they are *experts*) and that observer r gives, at the end of the experiment, value q_{ri} to sequence σ_i (and value q'_{rj} to sequence σ'_j). Then, we must perform a statistical test to detect bad observers in the panel (case of random observers); in words, a bad observer is one that does not (statistically) agree with the majority[3]. Assume (after a re-ordering of the observers' indexes) that observers $R' + 1, \cdots, R$ are bad ones. Then, their scores are taken out of the set, and the sequences receive as MOS

---

[b]Actually, we must do the whole process with several different sequences; we just present the technique for one of them.

the average of the values given to it by the good observers; that is, $Q_i = \sum_{r=1}^{R'} q_{ri}/R'$ (and $Q'_i = \sum_{r=1}^{R'} q'_{ri}/R'$).

(7) We now identify configuration γ_i with quality value Q_i (and γ'_j with Q'_j), and look for a real function $\nu()$ of 4 variables associated with the four selected parameters, such that for any set of values in the set $\mathcal{C}$ the function returns a number close to the associated MOS value. That is, for any $\gamma_i \in \mathcal{C}$, $\nu(\gamma_i) \approx Q_i$. This is the *learning* phase. It can be done with different tools. We tried standard Artificial Neural Networks (ANN), Bayesian Networks and RNN. As illustrated in Mohamed and Rubino[3], RNN performed by far the best, and that is the main reason why we used that tool.

Remark: before the learning phase, it is more comfortable to *scale* the input variables by dividing them by their maximal possible value (recall that in the second step we associate a range with each variable). This way, all input variables are in $[0..1]$; the same is done with the output, which corresponds to the normalised quality value.

(8) After having found the $\nu()$ function, we must go through the *validation* phase consisting of testing its value on the set of configurations in $\mathcal{C}'$. If for $\gamma'_i \in \mathcal{C}'$ it is $\nu(\gamma'_i) \approx Q'_i$, then $\nu()$ is validated and the process is finished.

Of course, if this is not the case, something was wrong in the previous process (not enough data, bad sampling of configurations, etc.).

(9) Using the evaluator consists now of measuring BR, FR, LR and MLBS, for instance at the receiver of a communication, and calling $\nu()$ with the measured values as input. This can be done in real time since the inputs can be collected in real time, and the evaluation of $\nu()$ is not expensive.

The next section briefly explains what is a RNN and how it is used in learning (phase (7)).

14.3. The Random Neural Networks Tool

We first present the mathematical RNN model and the particular case we use in the application presented in this chapter. Then, some elements about the use of the model as a learning tool are given. To have an idea about the multiple applications of this tool that have been already published[10].

14.3.1. *G-networks*

A Random Neural Network is a queuing network (also called a *G*-network), invented by E. Gelenbe in a series of papers[7–9] by merging concepts from neural networks and queuing theory. Its nodes are called queues or neurons, depending on the context, but they are the same mathematical object. The customers or units of the queuing network are called "signals" at the neural side. In this chapter, we will use both terminologies, the neural one and the concepts from queuing models, depending on the specific topic we are speaking about.

The set of nodes in the network is $\mathcal{N} = \{1, \cdots, N\}$. The queuing network receives two types of customers from outside, called *positive* and *negative*. Neuron i receives positive units from outside with rate $\lambda_i^+ \geq 0$ and negative units from outside with rate $\lambda_i^- \geq 0$. Both arrival processes are Poisson. At least one neuron receives positive units from outside; that is, $\sum_{i \in \mathcal{N}} \lambda_i^+ > 0$.

Nodes are FIFO queues (this is not essential but it simplifies the presentation) with a single exponential server and infinite room. The service rate at neuron i is $\mu_i > 0$. Positive units behave as usual customers in queuing networks: they arrive at the node and wait in the queue until the server is available, then get service and are sent to another queue or to outside. When a negative customer arrives at a queue, it destroys itself instantaneously, and if the queue was not empty, the last customer in it is also destroyed. Moving from a queue to another or to outside is an instantaneous process, as usual.

Observe that the previous description means that negative units can not be observed, only their effects can; the network never has negative customers in it. Denote by $X_i(t)$ the number of (positive) units in neuron i at time t, also called the *potential* of neuron i. Neuron i is said to be *active* at time t iff $X_i(t) > 0$. When neuron i is active, it sends units (positive or negative) to other neurons or to outside (with rate μ_i). When a (positive) customer ends getting service at neuron i, it goes to neuron j as a positive one with (routing) probability $p_{i,j}^+$ and as a negative one with (routing) probability $p_{i,j}^-$. The unit leaves the network with probability $d_i = 1 - \sum_{j \in \mathcal{N}} (p_{i,j}^+ + p_{i,j}^-)$.

The previous description means that when neuron i is active, it sends positive signals to neuron j with rate (also called here *weight*) $w_{i,j}^+ = \mu_i p_{i,j}^+$ and negative ones to neuron j with rate $w_{i,j}^- = \mu_i p_{i,j}^-$; it sends units outside with rate $\delta_i = \mu_i d_i$. Observe that $\delta_i + \sum_{j \in \mathcal{N}} (w_{i,j}^+ + w_{i,j}^-) = \mu_i$.

In the stable case, the *activity rate* of neuron i (the *utilisation rate* of queue i) is $\varrho_i = \lim_{t\to\infty} \Pr(X_i(t) > 0) > 0$; also, the mean throughput of positive units or signals that arrive at neuron i is $T_i^+ = \lambda_i^+ + \sum_{j\in\mathcal{N}} \varrho_j w_{j,i}^+$, and the mean arrival throughput of negative units at i is $T_i^- = \lambda_i^- + \sum_{j\in\mathcal{N}} \varrho_j w_{j,i}^-$. Look at neuron i as a queue and assume process $X_i()$ is stationary. Consider that destroyed customers are departures and apply the mean flow conservation theorem: we obtain $T_i^+ = \varrho_i(\mu_i + T_i^-)$ and thus,

$$\varrho_i = \frac{T_i^+}{\mu_i + T_i^-}.$$

Last, we place ourselves in the standard independence conditions concerning all arrival, service and switching (choosing next node and class, or output, for a signal leaving a neuron) processes (as usual in queuing network models). A basic result is then the following.

Theorem 14.1: The vector occupation process of this network is a Markov chain with state space $\mathbb{N}^N$. Assume it is irreducible (this depends on the routing probabilities of the model and on the arrival rates), and consider the relations

$$T_i^+ = \lambda_i^+ + \sum_{j\in\mathcal{N}} \varrho_j w_{j,i}^+, \quad T_i^- = \lambda_i^- + \sum_{j\in\mathcal{N}} \varrho_j w_{j,i}^- \quad \text{and} \quad \varrho_i = \frac{T_i^+}{\mu_i + T_i^-}$$

as a (non-linear) system of equations in the set of unknowns $(T_i^+, T_i^-, \varrho_i)_{i=1,\cdots,N}$. Then,

(i) The network is stable iff the system of equations has a solution where for $i = 1, \cdots, N$ we have $\varrho_i < 1$; in this case, the solution is unique.

(ii) In the stable case, the network is of the product-form type, and we have

$$\lim_{t\to\infty} \Pr(X_1(t) = n_1, \cdots, X_N(t) = n_N) = \prod_{i=1}^{N} (1 - \varrho_i)\varrho_i^{n_i}.$$

Proof: The proof is based on standard properties of Markov chains. See papers[7,8]. □

For instance, consider a single G-queue with arrival rate of positive (resp. negative) customers equal to $\lambda^+ > 0$ (resp. to $\lambda^- \geq 0$) and service rate $\mu > 0$. Then, applying Theorem 14.1 we have that the queue is stable iff $\lambda^+ < \mu + \lambda^-$, and in that case, its stationary state distribution is $n \mapsto (1-\varrho)\varrho^n$, where $\varrho = \lambda^+/(\mu+\lambda^-)$. The mean backlog at the queue (the mean potential of the neuron) is $\lambda^+/(\mu+\lambda^--\lambda^+)$. Observe that its occuppation process is that of a $M/M/1$ queue with arrival rate λ^+ and service rate $\mu+\lambda^-$.

14.3.2. Feedforward 3-layer G-networks

In a feedforward queuing network, a customer never visits the same node twice. This makes the analysis simpler. In the case of a G-network, the non-linear system presented in Theorem 14.1 has an "explicit" solution.

A 3-layer model is a particular case of feedforward networks. There are three types of nodes: the "input" nodes (their set is $\mathcal{I}$), which are the only ones receiving signals from outside, the "hidden" ones (their set is $\mathcal{H}$) receiving all the signals leaving input nodes, and the "ouput" nodes (their set is $\mathcal{O}$), receiving all the signals leaving hidden ones, and sending all their signals outside. There is no connection between nodes in the same subset. Moreover, for learning applications, it is usually assumed that there is no negative unit coming to the network from outside; we make the same assumption here.

The particular topology of the network makes that we have explicit expressions of the activity rates of all neurons, starting from set $\mathcal{I}$, then going to $\mathcal{H}$ and last to $\mathcal{O}$. Consider an input neuron i. Since no other neuron can send units to it, we have $\varrho_i = \lambda_i^+/\mu_i$. Now, for a hidden neuron h, we have

$$\varrho_h = \frac{\sum_{i\in\mathcal{I}} \varrho_i w_{ih}^+}{\mu_h + \sum_{i\in\mathcal{I}} \varrho_i w_{ih}^-} = \frac{\sum_{i\in\mathcal{I}} \lambda_i^+ w_{ih}^+/\mu_i}{\mu_h + \sum_{i\in\mathcal{I}} \lambda_i^+ w_{ih}^-/\mu_i}.$$

Last, for any output neuron o, the corresponding expression is

$$\varrho_o = \frac{\sum_{h\in\mathcal{H}} \varrho_h w_{ho}^+}{\mu_o + \sum_{h\in\mathcal{H}} \varrho_h w_{ho}^-} = \frac{\sum_{h\in\mathcal{H}} \frac{\sum_{i\in\mathcal{I}} \lambda_i^+ w_{ih}^+/\mu_i}{\mu_h + \sum_{i\in\mathcal{I}} \lambda_i^+ w_{ih}^-/\mu_i} w_{ho}^+}{\mu_o + \sum_{h\in\mathcal{H}} \frac{\sum_{i\in\mathcal{I}} \lambda_i^+ w_{ih}^+/\mu_i}{\mu_h + \sum_{i\in\mathcal{I}} \lambda_i^+ w_{ih}^-/\mu_i} w_{ho}^-}.$$

As we will see next, when this queuing network is used to learn, it is seen as a function mapping the rates of the arrival processes into the activity rates of the nodes. In this 3-layer structure, we see that the function

$\vec{\lambda}^+ = (\lambda_1^+, \cdots, \lambda_N^+) \mapsto \varrho_o$ is a rational one, and it can be easily checked that both its numerator and denominator are in general polynomials with degree $2H$ and non-negative coefficients.

14.3.3. *Learning*

Assume we are interested in some real function $f()$ from $[0..1]^{2N}$ into $[0..1]^N$. Function $f()$ itself is unknown, but we have K input-output values whose set is $\mathcal{D} = (\vec{l}^{(1)}, \vec{c}^{(1)}), \cdots, (\vec{l}^{(K)}, \vec{c}^{(K)})$ (that is, $f(\vec{l}^{(k)}) = \vec{c}^{(k)}$, $k = 1, \cdots, K$). We can use a stable G-network to "learn" $f()$ from the data set $\mathcal{D}$. For this purpose, consider a general RNN (not necessarily fedforward) with N nodes and all arrival rates in $[0..1]$. Now look at it as a black-box mapping the rates of the arrival process (the $2N$ numbers $\lambda_1^+, \cdots, \lambda_N^+, \lambda_1^-, \cdots, \lambda_N^-$) into the activity rates (the N numbers $\varrho_1, \cdots, \varrho_N$). We associate, for instance, the first N input variables of $f()$ with the arrival rates of positive units to nodes 1 to N, then the next N input variables with the arrival rates of negative units, last the N output variables of $f()$ with the activity rates of the G-network. The service rates and the routing probabilities (or equivalently, the weights) are seen as *parameters* of the mapping. The latter is here denoted as $\nu(\vec{\lambda})$, or $\vec{\nu}(\mathbf{W}, \vec{\mu}; \vec{\lambda})$, etc., depending on making explicit the parameters or not, with $\mathbf{W} = (W^+, W^-)$, $W^+ = (w_{ij}^+)$, $W^- = (w_{ij}^-)$, $\vec{\lambda} = (\vec{\lambda}^+, \vec{\lambda}^-)$, $\vec{\lambda}^+ = (\lambda_1^+, \cdots, \lambda_N^+)$, $\vec{\lambda}^- = (\lambda_1^-, \cdots, \lambda_N^-)$, and $\vec{\mu} = (\mu_1, \cdots, \mu_N)$.

Learning means looking for values of the parameters such that (i) for all $k = 1, \cdots, K$ we have $\nu(\vec{l}^{(k)}) \approx \vec{c}^{(k)}$ and, moreover, (ii) *for any other value* $\vec{x} \in [0..1]^{2N}$ we have $\nu(\vec{x}) \approx f(\vec{x})$. This last condition is experimentally validated, as usual in statistical learning, even if there are theoretical results about the way these $\nu()$ functions can approximate as close as desired any $f()$ function having some regularity properties. In order to find an appropriate G-network, the standard approach is to look only for network weights (matrix $\mathbf{W}$), the rest of the parameters being fixed. To do this, we consider the cost function

$$C(\mathbf{W}) = \frac{1}{2} \sum_{k=1}^{K} \sum_{o \in \mathcal{O}} \left[\nu_o(\mathbf{W}; \vec{l}^{(k)}) - c_o^{(k)} \right]^2,$$

and we look for minima of $C()$ in the set $\{\mathbf{W} \geq 0\}^c$. A basic way to find

---

[c] There can be stability issues here, depending on how we deal with the remaining parameters (service rates or departure probabilities), but we do not develop the point further for lack of space. Just observe that fixing the service rates high enough allows to control stability. For instance, $\mu_i = N$ for all neuron i is trivially sufficient for stability (recall we assume all arrival rates in $[0..1]$).

good values of $\mathbf{W}$ is to follow a gradient descent approach. This process helps in understanding the algebraic manipulations of these models while much more efficient procedures exist (for instance, quasi–Newton methods). For the basic approach, we build a sequence of matrices $\mathbf{W}(0)$, $\mathbf{W}(1)$, ..., (hopefully) converging to a pseudo-optimal one (that is, a matrix $\widehat{\mathbf{W}}$ such that $C(\widehat{\mathbf{W}}) \approx 0$). For instance, for all $i, j \in \mathcal{N}$, the typical update expressions for the weights between neurons i and j are

$$w_{i,j}^+(m+1) = w_{i,j}^+(m) - \eta \frac{\partial C}{\partial w_{i,j}^+}(\mathbf{W}(m)),$$

$$w_{i,j}^-(m+1) = w_{i,j}^-(m) - \eta \frac{\partial C}{\partial w_{i,j}^-}(\mathbf{W}(m)).$$

Factor $\eta > 0$ is called the *learning factor*; it allows to tune the convergence process. Observe that the previous expression can lead to a negative value for $w_{i,j}^+(m+1)$ or for $w_{i,j}^-(m+1)$. An usual solution to this is to use the iterations

$$w_{i,j}^+(m+1) = \left[w_{i,j}^+(m) - \eta \frac{\partial C}{\partial w_{i,j}^+}(\mathbf{W}(m)) \right] \vee 0,$$

$$w_{i,j}^-(m+1) = \left[w_{i,j}^-(m) - \eta \frac{\partial C}{\partial w_{i,j}^-}(\mathbf{W}(m)) \right] \vee 0.$$

Another possible decision is to *freeze* the value of an element of $\mathbf{W}(m)$ when it reaches value zero. These are standard points in basic optimisation methodology.

It remains how to compute the partial derivative in the previous expression. Writing

$$\frac{\partial C}{\partial w_*^*} = \sum_{k=1}^{K} \sum_{o \in \mathcal{O}} \left[\nu_o(\mathbf{W}; \vec{\imath}^{(k)}) - c_o^{(k)} \right] \frac{\partial \nu_o}{\partial w_*^*},$$

we see that we still need to calculate the partial derivatives of the outputs, that is, of the activity rates, with respect to the weights.

After some algebra we can write the following relations: if $\vec{\varrho}$ is the vector $\vec{\varrho} = (\varrho_1, \cdots, \varrho_N)^{\mathrm{d}}$,

$$\frac{\partial \vec{\varrho}}{w_{uv}^+} = \vec{\gamma}_{uv}^+ (\mathbf{I} - \mathbf{\Omega})^{-1}, \quad \frac{\partial \vec{\varrho}}{w_{uv}^-} = \vec{\gamma}_{uv}^- (\mathbf{I} - \mathbf{\Omega})^{-1},$$

---

[d] All vectors are row vectors in this chapter.

where
$$\Omega_{jk} = \frac{w_{jk}^+ - \varrho_k w_{jk}^-}{\mu_k + T_k^-},$$

$$\vec{\gamma}_{uv}^+ = \frac{\varrho_u}{\mu_v + T_v^-}\vec{1}_v, \quad \vec{\gamma}_{uv}^- = -\frac{\varrho_u \varrho_v}{\mu_v + T_v^-}\vec{1}_v = -\varrho_v \vec{\gamma}_{uv}^+,$$

vector $\vec{1}_j$ being the jth vector of the canonical base of $\mathbb{R}^N$.

A compact way of writing these derivatives is as follows. Denote by $\mathbf{R}$ the diagonal matrix $\mathbf{R} = \mathrm{diag}(\varrho_i)_{i \in \mathcal{N}}$ and by $\mathbf{D}$ the diagonal matrix $\mathbf{D} = \mathrm{diag}(\mu_i + T_i^-)_{i \in \mathcal{N}}$. Now, let i be fixed and let $A^{(i)} = (A_{uv}^{(i)})$ and $B^{(i)} = (B_{uv}^{(i)})$ be given by

$$A_{uv}^{(i)} = \frac{\partial \varrho_i}{w_{uv}^+}, \quad B_{uv}^{(i)} = \frac{\partial \varrho_i}{w_{uv}^-}.$$

We have:
$$A_{uv}^{(i)} = \frac{\varrho_u M_{vi}}{\mu_v + T_v^-} = \frac{\varrho_u M_{vi}}{D_{vv}}, \quad B_{uv}^{(i)} = -\frac{\varrho_u \varrho_v M_{vi}}{D_{vv}} = -A_{uv}^{(i)},$$

where $\mathbf{M} = (\mathbf{I} - \mathbf{\Omega})^{-1}$. Vector $(M_{1i}, \cdots, M_{Ni})^{\mathrm{T}}$, the ith column of $\mathbf{M}$, can be written $\mathbf{M}\vec{1}_i^{\mathrm{T}}$. We now can write

$$A^{(i)} = \vec{\varrho}^{\mathrm{T}} \left(\mathbf{M}\vec{1}_i^{\mathrm{T}}\right)^{\mathrm{T}} \mathbf{D}^{-1} = \vec{\varrho}^{\mathrm{T}} \vec{1}_i \mathbf{M}^{\mathrm{T}} \mathbf{D}^{-1}.$$

In the same way,
$$B_{uv}^{(i)} = -\vec{\varrho}^{\mathrm{T}} \vec{1}_i \mathbf{M}^{\mathrm{T}} \mathbf{D}^{-1} \mathbf{R}.$$

Resuming, learning means minimising and minimising means in this context, computing derivatives. This in turn means, in the general case, inverting a matrix whose dimension is equal to the number of parameters (the variables in the minimisation process). In the particular case of feedforward models, the inversion can be done very easily (with an appropriate order in the variables, the matrix to invert is a triangular one).

14.3.4. Sensitivity Analysis

One of the consequences of the nice mathematical properties of G-networks is that they allow to perform sensitivity analysis in a systematic way. Sensitivity analysis means here to be able to compute the derivatives of the activity rates with respect to the arrival rates.

After some straightforward algebra, we get the following relations: if $u \neq i$,

$$\frac{\partial \varrho_i}{\partial \lambda_u^+} = \sum_j \frac{\partial \varrho_j}{\partial \lambda_u^+} \Omega_{ji},$$

and

$$\frac{\partial \varrho_i}{\partial \lambda_i^+} = \sum_j \frac{\partial \varrho_j}{\partial \lambda_i^+} \Omega_{ji} + \frac{1}{\mu_i + T_i^-}.$$

In the same way, if $u \neq i$,

$$\frac{\partial \varrho_i}{\partial \lambda_u^-} = \sum_j \frac{\partial \varrho_j}{\partial \lambda_u^-} \Omega_{ji},$$

and

$$\frac{\partial \varrho_i}{\partial \lambda_i^-} = \sum_j \frac{\partial \varrho_j}{\partial \lambda_i^-} \Omega_{ji} - \frac{\varrho_i}{\mu_i + T_i^-}.$$

In matrix form, let us denote $\mathbf{F} = (F_{iu})$ and $\mathbf{G} = (G_{iu})$ where $F_{iu} = \partial \varrho_i / \partial \lambda_u^+$ and $G_{iu} = \partial \varrho_i / \partial \lambda_u^-$. If $u \neq i$, we have

$$F_{iu} = \sum_j F_{ju} \Omega_{ji} = \sum_j \Omega_{ij}^{\mathrm{T}} F_{ju} \quad \text{and} \quad G_{iu} = \sum_j G_{ju} \Omega_{ji} = \sum_j \Omega_{ij}^{\mathrm{T}} G_{ju}.$$

When $u = i$,

$$F_{ii} = \sum_j \Omega_{ij}^{\mathrm{T}} F_{ju} + \frac{1}{\mu_i + T_i^-} \quad \text{and} \quad G_{ii} = \sum_j \Omega_{ij}^{\mathrm{T}} G_{ju} - \frac{\varrho_i}{\mu_i + T_i^-}.$$

Using the same notation than in previous subsection, this leads to the expressions

$$\mathbf{F} = \mathbf{D}^{-1} \mathbf{M}^{\mathrm{T}} \quad \text{and} \quad \mathbf{G} = \mathbf{D}^{-1} \mathbf{R} \mathbf{M}^{\mathrm{T}}.$$

As we see, the general formulæ for performing sentitivity analysis in the general case need the inversion of the same matrix as for the learning process. As before, a feedforward network structure simplifies considerably the computations (only a triangular matrix must be inverted).

14.4. Applications

In the case of the example described in Section 14.2, we use a RNN having 4 input nodes (corresponding to the 4 selected variables, BR, FR, LR and MLBS) and one output node, corresponding to the quality of the flow. Some details about the performance of RNN in the learning phase, their relative insensitivity with respect to the size of the subset of hidden nodes (in a reasonable range), and in general, their main properties, can be seen in Mohamed and Rubino[3]. For audio flows, see Mohamed, Rubino and Varela[4].

The first direct application of the PSQA technology is to analyse the impact of the selected variables on perceived quality. Since once the RNN has learnt from data we have a function giving quality values for any configuration in the space we defined at the beginning of the process, we are able to analyse this quality function as a function of a part of the input vector, to compute its sensitivity with respect to the input variables, etc. Again, we refer to the given papers to see details about these results.

A second and richer set of results is as follows. Suppose you are interested in analysing the impact of some specific part of the communication structure on quality. For instance, consider again our video example, and assume you want to analyse the impact of the size H of a specific buffer on quality. If you have a model that is able to evaluate the impact of H on the network parameters you selected when applying PSQA (LR and MLBS in our example), then you can map H into quality and perform your analysis efficiently. An example of this is Rubino and Varela[6], where we analyse the effect of FEC (Forward Error Correction) on the quality of voice flows. The basis of the approach are in Rubino and Varela[5]. Let us explain here this approach with more details, but keeping our video example of Section 14.2.

Assume we send a video stream using packets of constant length B bits through the Internet, from some source S to a receiver R. Applying the PSQA methodology, we get an explicit function $Q = \nu(\text{BR}, \text{FR}, \text{LR}, \text{MLBS})$ where BR is the bit rate of the connection, FR its frame rate, LR is its end-to-end loss probability and MLBS the average size of its bursts of losses.

The connection has a bottleneck with capacity c bps, and it is able to store up to H packets. For simplicity, assume that a packet contains exactly one frame. The flow of packets arriving to this bottleneck is Poisson with rate λ pps (packets per second). With the standard assumptions, this node is a $M/M/1/H$ queue (at the packet level), leading to simple expressions of the loss probability p and (see Rubino and Varela[5]) mean loss burst size M. We have $p = (1-\varrho)\varrho^H/(1-\varrho^{H+1})$ and, after analysing the Markov chain, $M = 1 + \varrho$, where the load $\varrho = \lambda B/c$ is assumed to be $\neq 1$. If our flow is the only one using the bottleneck node, its quality can be written $Q = \nu(\lambda B, \lambda, p, M)$. More explicitly, this gives

$$Q = \nu\left(\lambda B, \lambda, \frac{(1-\lambda B/c)(\lambda B/c)^H}{1-(\lambda B/c)^{H+1}}, 1+\lambda B/c\right).$$

If other flows share the same bottleneck, then the difficulty lies in analysing the loss rate and the mean loss burst size of our connection at that node. For instance, consider the case of K different flows sharing the node, with troughputs $\lambda_1, \cdots, \lambda_K$. Our flow is number 1. For simplicity, let us consider that all packets sizes are the same, B bits, and that the node can handle up to H packets of all types. With the standard exponential assumptions, we have now a multiclass FIFO $M/M/1/H$ model with input rates $\lambda_1, \cdots, \lambda_K$ and service rate $\mu = c/B$. We need now to compute the loss probability for class 1 customers and the mean loss burst size of these customers. Since we assume Poisson arrivals, the loss probability is the same for all classes and thus can be computed from the $M/M/1/H$ single class formula with $\lambda = \lambda_1 + \cdots + \lambda_K$. It remains the problem of the mean burst loss size for class 1 customers. The first sub-problem is to define a burst in this multiclass context. Following Rubino and Varela[6], we define a burst in the following way. Assume a packet of class 1 arrives at a full queue and is lost, and assume that the previous class 1 packet that arrived was not loss. Then, a burst of class 1 losses starts. The burst ends when a packet is accepted, whatever its class is. See Rubino and Varela[6] for a discussion about this definition.

Let us denote by LBS the loss burst size for class 1 units. We have that $\Pr(\text{LBS} > n) = q^n$, where

$$q = \sum_{m \geq 0} \left(\frac{\lambda - \lambda_1}{\lambda + \mu}\right)^m \frac{\lambda_1}{\lambda + \mu} = \frac{\lambda_1}{\lambda_1 + \mu}.$$

The last relationship comes from the fact that we allow any number of class $k \neq 1$ units to arrive as far as their are lost, between two class 1 losses (that is, while the burst is not finished, no departure from the queue is "allowed"). Using the value of q, we have

$$\text{E(LBS)} = \sum_{n \geq 0} \Pr(\text{LBS} > n) = \cdots = 1 + \frac{\lambda_1}{\mu}.$$

With a straightforward extension of the results in Marie and Rubino[11], the analysis can be extended to the case of packets with different sizes for different flows. The discussion's aim was to show the kind of difficulty we may have to face at, when coupling PSQA to standard performance models. In some more complex cases the mapping from the variables the users wants to analyse and the inputs to the $\nu()$ function will need a simulator.

14.5. Conclusions

Using RNN, we have developed a technology allowing to put the *perceived* quality of a video (or audio, or multimedia) stream (that is, a subjective object, by definition) into numbers, after the flow arrives at the receiver through the Internet, and *automatically*. This evaluation process is not time consuming and can therefore be done *in real time*, if necessary.

When coupled with a standard performance model, this approach can allow the analyst to dimension a system adressing directly quality and not indirected metrics (such as those related to losses, delays, etc.). For instance, he/she will find the optimal channel speed (or buffer size, or number of terminals) that allows keeping quality over some threshold, instead of doing the same with loss rates, or delays, or jitter. The approach allows to work directly with quality, the "ultimate target".

Ongoing work with this technique includes applying it to control experiments, or to quality assessment in an operating network. Concerning the methodology itself and in particular the RNN tool, ongoing efforts are being done in order to explore efficient learning algorithms together with analysing the mathematical properties of these specific dynamic models.

References

1. ITU-T Recommendation P.800, "Methods for subjective determination of transmission quality" (http://www.itu.int/).
2. ITU-R Recommendation BT.500-10, "Methodology for the subjective assessment of the quality of television pictures" (http://www.itu.int/).
3. Mohamed, S. and Rubino, G., "A study of real–time packet video quality using Random Neural Networks" *IEEE Transactions On Circuits and Systems for Video Technology*, **12** (12), (Dec. 2002), 1071-1083.
4. Mohamed, S. and Rubino, G. and Varela, M., "Performance evaluation of real-time speech through a packet network: a Random Neural Networks-based approach" *Performance Evaluation*, **57**(2), (2004), 141-162.
5. Rubino, G. and Varela, M., "A new approach for the prediction of end-to-end performance of multimedia streams",IEEE CS Press, 2004, 110-119. (1st International Conference on Quantitative Evaluation of Systems (QEST'04),University of Twente, Enschede, The Netherlands , Sept. 2004)
6. Rubino, G. and Varela, M., "Evaluating the utility of media–dependent FEC in VoIP flows", in Quality of Service in the Emerging Networking Panorama, LNCS 3266, 2004, 31-43.
7. Gelenbe, E., "Random Neural Networks with negative and positive signals and product form solution" *Neural Computation*,**1**(4), (1989), 502–511.
8. Gelenbe, E., "Stability of the Random Neural Network model" ISBN 3540522557, 1990, 56-68. (Proc. of Neural Computation Workshop, Berlin, Germany, Feb. 1990).

9. Gelenbe, E., "Learning in the recurrent Random Neural Network" *Neural Computation*, **5**(1), (1993), 154-511.
10. Bakircioglu, H. and Kocak, T., "Survey of Random Neural Network applications" *European Journal of Operational Research*, **126**(2), (2000), 319-330.
11. Marie, R. and Rubino, G., "Semi-explicit formulas for the M/M/1 Multiclass Queue" *Journal of the Theoretical Computer Science Institut*, Polish Academy of Sciences, **2**(1-2), (1990), 7-18.

CHAPTER 15

A Study of the Dynamic Behaviour of a Web Site

Maria Carla Calzarossa

Dipartimento di Informatica e Sistemistica, Università di Pavia
via Ferrata 1, I – 27100 Pavia, Italy
E-mail: mcc@unipv.it

Daniele Tessera

Dipartimento di Matematica e Fisica, Università Cattolica del Sacro Cuore
via Musei 41, I – 25121 Brescia, Italy
E-mail: d.tessera@dmf.unicatt.it

Interactive Web services make use of highly dynamic contents. To design efficient mechanisms for the replication and distribution of these contents and improve the QoS perceived by the users, it is important to understand how often and to what extent contents change. This chapter addresses these issues by studying the dynamic behaviour of the contents of a popular news Web site. We analysed the contents of the news in terms of various metrics to assess the similarity of their successive versions. Moreover, we identified groups of news characterised by similar behaviour.

15.1. Introduction

The increased pervasiveness of interactive services offered over Internet opens new performance challenges. These services are typically used by a large number of users who share various types of contents that are updated as a consequence of external events or of actions performed by the users themselves. The peaks of load on the servers and the peaks of traffic over the network can cause delays in delivering the requested contents and propagating the updates. These delays have a negative impact on the QoS perceived by the users who might even end up by accessing out of date contents.

To cope with QoS requirements without overprovisioning the systems, solutions, such as, Content Distribution Networks, and peer-to-peer systems, have been adopted. These solutions allow the large scale replication of shared contents, hence, require efficient data distribution mechanisms.

The analysis of the update process of Web contents is the starting point of any study aimed at assessing the performance and scalability of the proposed solutions. These issues have been addressed from different perspectives[1-3,7,8,15] A combination of empirical data and analytic modeling is used to estimate the change rate of Web pages[2]. The stochastic properties of the dynamic page update patterns and their interactions with the corresponding request patterns are addressed by Challenger *et al.*[4] Their study shows that the pages of highly dynamic sport sites are characterised by relatively large bursts of updates and periodic behaviour. Cho and Garcia–Molina[7] introduce estimators to study the frequency of change to a data item (e.g., a Web page) in the presence of incomplete change history of the item itself. The process of creation and modification of HTML files of the MSNBC news Web site is addressed by Padmanabhan and Qiu[13]. Their study shows that files tend to change little when they are modified and modification events tend to concentrate on a small number of files. Metrics that consider the introduction of new files and their influence on the requests to the Web site are introduced by Cherkasova and Karlsson to study the dynamics and evolution of Web sites[6]. The impact of new files is also addressed in the framework of media sites[5]. Fetterly *et al.*[9] analyse the rate and degree of change of Web pages and the factors correlated with change intensity. The study shows that document size is strongly related to both frequency and degree of change. In particular, large documents change more often and more extensively than smaller documents.

This chapter studies the evolution of a news Web site with the objective of analysing how often and to what extent the contents of Web pages change. We chose the MSNBC Web site as we consider it a good representative of the news Web sites. Since we did not have access to any server log, we had to monitor the site. In this respect and many others, our study differs from a previous study on the same site[13]. In particular, we focused on the "core" of the HTML files, that is, the text of the news, and we studied the change frequency and the amount of change to each news between two successive downloads. We used various metrics to analyse the core of the news and assess the similarity of their successive versions. Moreover, by applying clustering techniques, we identified groups of news characterised by similar behaviour.

The chapter is organised as follows. Section 15.2 presents the experimental set up used for the data collection. Section 15.3 describes the results of our analysis. Finally, conclusions and future research directions are outlined in Sec. 15.4.

15.2. Data Collection

Our study relies on data collected from the MSNBC news Web site[12]. Since we were interested in the actual contents of Web pages, we resorted to active monitoring and we downloaded the HTML files from the site. From these repeated accesses we could detect changes to each file and estimate the change frequency and the extent of each change. Let us remark that by change we mean any modification to the file.

Our monitoring adopted a conservative approach in that at each access we downloaded all the HTML files that were on the site, including the set of files uploaded to the site since the last access. Note that we only downloaded the HTML files classified on the MSNBC Web site under five major categories, namely, news, business, sports, entertainment, health. On the average the number of new files downloaded over a 24 hour period is equal to 104.36, with a distribution characterised by low variability. Even though the number of new files can be as large as 156 and as small as 33, its standard deviation is equal to 29.93, that is, much lower than the corresponding mean.

To avoid an uncontrolled increase of the download time, we did some post-processing of the files as to identify the files still worth monitoring. In particular, we decided to filter out from the list of files to be downloaded, the files that were not active, that is, did not receive any update over a period of five consecutive days.

Let us remark that the size of the HTML files was not a good indicator of the actual updates, due to the presence of various types of contents, e.g., advertisement, that typically change at every access. Hence, we did parse the files to identify changes to their core. Note that at each access we downloaded on the average about 500 files.

News Web sites tend to update their contents periodically whenever something new happens. Hence, the granularity of the monitoring interval, that is, how often to download a particular file, was a crucial issue for the accuracy of our analysis. If a file changes once a hour, it might be unnecessary to download it every minute, whereas it might be insufficient to download it once a day. As a consequence, the granularity of the monitoring has to be fine enough to capture all updates and coarse enough to minimise the time spent to download all files over the network.

For this purpose, we monitored the MSNBC Web site over intervals of various lengths. Our analysis has shown that a granularity of 15 minutes is a good tradeoff between the frequency of changes and the download time. The measurements were then collected at regular intervals of 15 minutes from the middle of November 2004 for a period of approximately 16 weeks.

To remove the fluctuations due to the warm-up effects of the monitoring process, we discarded the first week of measurements. Similarly, we discarded the "tail" of the measurements. The total number of HTML files successfully downloaded was about 5 millions. Among them, the number of unique files was equal to 9,209. Note that each file was downloaded at least 480 times, that is, we collected at least 480 versions of the same file.

15.3. Results

As a preliminary exploratory analysis, we focused on the overall characteristics of the HTML files. Table 15.1 presents some descriptive statistics of the size of the HTML files, with a breakdown into the five news categories considered in our study. As can be seen, files tend to be rather small, their average size is about 42Kbytes, and there is little variation across categories. The table also shows the number of unique files downloaded for each category.

Table 15.1. Descriptive statistics of the size, expressed in bytes, of the HTML files.

Category	mean	standard dev.	min	max	unique files
News	40,812	4,321	13,895	83,980	3,393
Business	39,863	3,173	13,849	57,963	1,448
Sports	45,425	4,777	13,894	72,175	2,588
Entertainment	43,557	4,277	28,164	92,018	915
Health	39,636	4,028	14,295	59,942	865
Global	42,524	4,949	13,894	92,018	9,209

Since we were interested in analysing the core of the news, that is, their actual contents, we did some preprocessing of each HTML file to extract the text of the news and the HTML tags making up its layout. Moreover, as we did not download the objects, e.g., images, videos, embedded in the HTML files, we considered as part of the core of the news the HTML tags and URLs of all embedded objects, but advertisement. Indeed, we have seen that tags and URLs are good indicators of the contents of the corresponding objects. In what follows, our analysis then focuses on the core of the news.

The number of dynamic news, that is, news that received at least one update over their monitoring period, is equal to 4,252, and accounts for about the 46.2% of the total number of unique files downloaded during the entire monitoring interval. The remaining 4,957 news can be classified as static, in that they did not receive any update during their monitoring period. Note that the number of dynamic news is not evenly distributed across the five categories. For example, 39% of the dynamic news belong to the sports category, whereas less than 7.4% belong to the entertainment category. Moreover, 63.8% of the news within the sports category are dynamic. On the contrary, the percentage of dynamic news within the entertainment category is equal to 34.3.

Table 15.2 presents the statistics of the number of changes to the 4,252 dynamic news. As can be seen, this number varies across the categories. On the average, each news received 2.732 changes, even though there are news that received as many as 118 changes.

Table 15.2. Descriptive statistics of the number of changes to the dynamic news.

Category	mean	standard dev.	max	total number of changes
News	2.738	3.871	49	3,431
Business	2.398	2.905	48	1,698
Sports	3.077	4.575	118	5,084
Entertainment	2.650	2.918	22	832
Health	1.761	1.504	13	572
Global	2.732	3.859	118	11,617

The distribution of the intervals between two successive changes to a news is characterised by a mean value equal to 214.18 minutes and a standard deviation about four times larger. This distribution is highly positively skewed. Note that the time stamp associated to each news corresponds to the time measured when the first byte of the corresponding HTML file was received from the site.

Figure 15.1 shows the distribution of the changes detected on all news as a function of the time elapsed since the news was uploaded into the site. The figure plots the changes occurred within 72 hours, that account for 99.6% of the total number of changes. As can be seen, a large fraction of the changes, namely, 78%, occurs within 12 hours, and 97% within 48 hours. The distribution of number of changes varies across the five categories of news considered in our analysis. For example, 80% of changes to the health news occur within six hours, and about 98% within 12 hours. For the sports news, 76% of the changes occur within 12 hours and 96% within 48 hours.

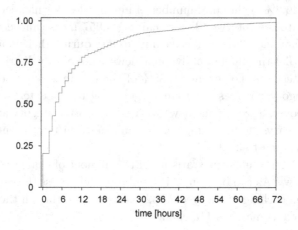

Fig. 15.1. Distribution of the number of changes as a function of time.

Similarly, our analysis has shown that the majority of the news, namely, 80%, received all their updates within 12 hours. This percentage goes up to 96% if we consider an interval of 48 hours.

To discover similarities in the characteristics of the news, we applied clustering techniques. Each news was described by three parameters, namely, number of changes, average time between two successive changes, and average size of the core of the news. Clustering yields a partition of the news into four groups. Table 15.3 presents the centroids, i.e., the geometric centers, of the groups. As can be seen, the news belonging to group 1, that accounts for about 62% of the news, are characterised by a small number of changes and by an interval between changes of approximately two hours. Group 4, the smallest group that accounts for the 7% of the news, contains the news characterised by a number of changes, namely, 11.587, more than four times larger than the global average of all news (equal to 2.732). The news with the largest size are grouped in cluster 2, whereas the news with the longest interval between successive changes belong to cluster 3.

The analysis of the composition of each cluster in terms of the categories of the news has shown that about 52% of the news of cluster 4 belong to the sports category, whereas the majority of the news of clusters 2 and 3

Table 15.3. Centroids of the four groups obtained for the dynamic news.

	number of changes	time between changes [min]	size [bytes]	fraction of news
Group 1	1.871	111.929	3,303	62%
Group 2	2.688	154.711	6,872	20%
Group 3	2.253	587.184	3,785	11%
Group 4	11.587	121.889	4,911	7%

belong to the news category. About 65% of the business news and 76% of the health news belong to cluster 1.

To better understand the dynamic behaviour of the news, we have analysed to what extent their contents change in terms of size. On the average, the size of the core of two successive versions of a news differs by 480.45 bytes. This value varies across the categories. For example, for the news category, the average difference is equal to 556.6 bytes. Moreover, the analysis has shown that, as expected, the size of successive versions tends to increase. To quantify the extent of the changes, we have also studied the relative variation of the size of the core of successive versions of each news, namely:

$$\left| \frac{s_j - s_{j+1}}{s_j} \right|$$

where s_j and s_{j+1} denote the size of the j-th and of the $(j+1)$-th version of a news, respectively. Figure 15.2 shows the distribution of the relative variation computed over all news belonging to the news category. The average variation is equal to 0.19. Note that this value corresponds to the 75-th percentile of the distribution. On the contrary, the median of the distribution is equal to 0.066. Let us remark that the figure shows the distribution in the range $[0, 2]$, that accounts for approximately 99% of the values.

To further analyse the amount of change to successive versions of a news, we represented the news through the words that they contained using the vector space model typical of information retrieval[14]. Let N denote the size of the vocabulary, that is, the global number of distinct words in all versions of a specific news. The j-th version of a news is then mapped into an N-dimensional space, namely, it is represented by a vector $d_j = (w_{1,j}, w_{2,j}, \ldots, w_{N,j})$, where $w_{i,j}$ denotes the weight of word i in the j-th version of the news. In our analysis, as weight we used the raw term frequency, that is, the number of times each word occurred in the core of the news. Hence, words that occur frequently in a news are more important than infrequent words. Moreover, despite information retrieval applications,

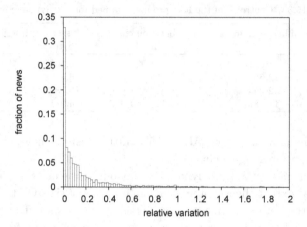

Fig. 15.2. Distribution of the relative variation of the size of successive versions of all news in the news category.

we considered all words including the so-called "stop words".

The similarity measures are based on the various metrics, e.g., cosine, Jaccard and Dice coefficients[10,11,16]. In our analysis we used the cosine metrics. According to this metrics, the similarity between two documents d_j and d_k is given by:

$$sim(d_j, d_k) = \frac{d_j \cdot d_k}{|d_j| \times |d_k|} = \frac{\sum_{i=1}^{N} w_{i,j} \times w_{i,k}}{\sqrt{\sum_{i=1}^{N} w_{i,j}^2} \times \sqrt{\sum_{i=1}^{N} w_{i,k}^2}}$$

The values of this metrics are in the range $[0, 1]$. The cosine coefficient is equal to 0 whenever there are no words in common between two successive versions of a news. On the contrary, it is equal to 1 for two identical versions, that is, with the same word frequency.

Figure 15.3 shows the distribution of the cosine coefficient of similarity computed for all the dynamic news considered in our analysis. The values refer to the similarity between pairs of successive versions of the news. The figure plots the distribution in the range $[0.6, 1]$, that accounts for 99.4% of the values. As can be seen, the distribution is skewed towards 1. Its average is equal to 0.96125. This means that in terms of words, successive versions of the news are rather similar. It is interesting to point out that the minimum of the distribution, that is equal to 0.02916, corresponds to news belonging to the business category. Similarly, the minimum values of the cosine coefficient computed for the news belonging to the news and

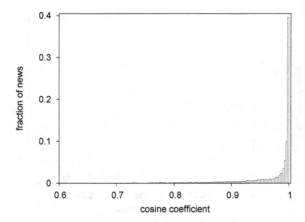

Fig. 15.3. Distribution of the cosine coefficient of similarity computed between pairs of successive versions of each individual news.

health categories are rather small, i.e., 0.03275 and 0.03469, respectively. On the contrary, the minimum values for the sports and entertainment news are about one order of magnitude bigger, namely, 0.31885 and 0.57093, respectively.

Another application of the cosine coefficient of similarity was aimed at testing to what extent a news changes with respect to its first version, that is, which version is the closest to the first version. For this purpose, we computed the cosine coefficients between the first version of a news and all its successive versions. The average value of the cosine coefficient is equal to 0.84887, that is, much lower than the value previously computed for the pairs of the successive versions of each news. Figure 15.4 shows the distribution of the cosine coefficient of similarity for the news belonging to the news category. As can be seen, successive versions tend to differ significantly from the first version of the news. The average coefficient of similarity is equal to 0.80038.

15.4. Conclusions

Dynamics of Web contents have to be taken into account when making a decision about caching, content distribution and replication. We studied the evolution of a popular news Web site with the objective of understanding how often and to what extent its contents change. Despite what expected, our exploratory analysis has shown that the core of most of the news tends

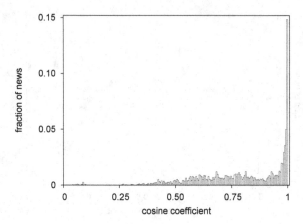

Fig. 15.4. Distribution of the cosine coefficient of similarity for the news category computed between the first version of a news and all its successive versions.

to change little and not very often, even though this behaviour varies across news categories. For example, sports news change more frequently. On the contrary, news belonging to the news category tend to change to a larger extent. Similarity measures have also shown that successive versions of the same news do not differ significantly, whereas the difference increases when we compare the first version with all its successive versions.

As a future work, we plan to derive models able to capture and reproduce the dynamic behaviour and the evolution of the contents of the Web site. Moreover, we plan study the performance implications of this behaviour.

Acknowledgments

This work was supported by the Italian Ministry of Education, Universities and Research under the FIRB programme. Authors wish to thank Clara Parisi for her valuable help in setting up the experimental environment.

References

1. Barili, A., Calzarossa, M.C. and Tessera, D., "Characterization of Dynamic Web Contents" LNCS 3280 - Springer, 2004, 648-656. (In Computer and Information Sciences - ISCIS 2004).
2. Brewington, B.E. and Cybenko, G., "How Dynamic is the Web?" *Computer Networks*, **33**(1-6), (2000), 257-276.
3. Brewington, B.E. and Cybenko, G., "Keeping Up with the Changing Web". *IEEE Computer*, **33**(5), (2000), 52-58.

4. Challenger, J.R., Dantzig, P., Iyengar, A., Squillante, M.S. and Zhang, L., "Efficiently Serving Dynamic Data at Highly Accessed Web Sites". *IEEE/ACM Transactions on Networking*, **12**(2), (2004), 233-246.
5. Cherkasova, L. and Gupta, M., "Analysis of Enterprise Media Server Workloads: Access Patterns, Locality, Content Evolution, and Rates of Change". *IEEE/ACM Transactions on Networking*, **12**(5), 2004, 781-794.
6. Cherkasova, L. and Karlsson, M., "Dynamics and Evolution of Web Sites: Analysis, Metrics and Design Issues", 2001, 64-71. (In Proc. of the 6th IEEE Symposium on Computers and Communications, 2001).
7. Cho, J. and Garcia-Molina, H., "Estimating Frequency of Change". *ACM Transactions on Internet Technology*, **3**(3), 2003, 256-290.
8. Douglis, F., Feldmann, A., Krishnamurthy, B. and Mogul, J., "Rate of Change and other Metrics: A Live Study of the World Wide Web", 1997, 147-158. (In Proc. of the First USENIX Symposium on Internet Technologies and Systems, 1997).
9. Fetterly, D., Manasse, M., Najork, M. and Wiener, J., "A Large-Scale Study of the Evolution of Web Pages". *Software - Practice and Experience*, **34**(2), (2004), 213-237.
10. Lee, D.L., Chuang, H. and Seamons, K., "Document Ranking and the Vector-Space Model". *IEEE Software*, **14**(2), (1997), 67-75.
11. Lee, L., "Measures of Distributional Similarity", 1999, 25-32. (In Proc. of the 37th Meeting of Association for Computational Linguistics, 1999).
12. "MSNBC Web site". http://www.msnbc.com.
13. Padmanabhan, V.N. and Qiu, L., "The Content and Access Dynamics of a Busy Web Site: Findings and Implications", 2000, 111-123. (In Proc. of the ACM SIGCOMM Conference on Applications, Technologies, Architectures, and Protocols for Computer Communication, 2000).
14. Salton, G. and McGill, M.J., *Introduction to Modern Information Retrieval*. (McGraw-Hill, New York), 1983.
15. Shi, W., Collins, E. and Karamcheti, V., "Modeling Object Characteristics of Dynamic Web Content". *Journal of Parallel and Distributed Computing*, **63**(10), (2003), 963-980.
16. Zobel, J. and Moffat, A., "Exploring the Similarity Space". *ACM SIGIR Forum*, **32**(1), (1998), 18-34.